Damselflies

of Minnesota, Wisconsin *&* Michigan

Robert DuBois

Kollath+Stensaas
PUBLISHING

Kollath+Stensaas Publishing
394 Lake Avenue South, Suite 406
Duluth, MN 55802
Office: 218.727.1731
Orders: 1.800.678.7006 (Adventure Keen Distributors)
info@kollathstensaas.com
www.kollathstensaas.com

DAMSELFLIES *of* MINNESOTA,
WISCONSIN & MICHIGAN

Printed in Duluth, Minnesota U.S.A. by J.S. Print Group
10 9 8 7 6 5 4 3 2 1 First Edition

Editorial Director: Mark Sparky Stensaas
Graphic Designer: Rick Kollath

ISBN-13: 978-1-936571-13-0

Table of Contents

To my wife Linda,
and daughters
Danielle, Christina & Sarah

And to my parents,
Robert P. and Malvina

And to my many friends within the community of odonate enthusiasts who made chasing damselflies fun

Acknowledgements

I could not have written this book without the support of my family First, I thank my wife Linda, who has been a continual source of optimism and encouragement, and who has helped create time and space for me to pursue my passion for all things odonatological.

My parents, Robert P. and Malvina DuBois, have enthusiastically supported and encouraged my interest in the natural world since I was a kid. And to my daughters Danielle, Christina, and Sarah: thank you for joining me on collecting trips, for patiently listening to far too many lengthy discourses on every conceivable topic related to the Odonata, and for encouraging me every step of the way.

I also want to acknowledge the help I've received in studying Odonata in the Upper Midwest, and the friendships that have developed, from the following odonate enthusiasts: Matt Berg, Ryan Brady, Ryan Chrouser, Marla Garrison, Mitch Haag, Dan Jackson, Denny Johnson, Colin Jones, Scott King, Joanne Kline, Ron Lawrenz, Karl Legler, John and Cindy McKee, Kurt Mead, Mark O'Brien, Curt Oien, Julie Pleski, Mike Reese, Kurt Schmude, Wayne Steffens, Bill Smith, Ken Tennessen, Ami Thompson, and Freda van den Broek.

I was fortunate to cross paths with an illustrator with the skills of Rick Kollath, and a layout designer with the vision of Sparky Stensaas; thank you for your hard work and patience.

I'm appreciative as well of the many folks who contributed their fine photographs for this book; Mike Reese, Sparky Stensaas, Ken Tennessen, Sid Dunkle, Dennis Paulson, Mike Furtman, Dan Jackson and Kurt Huebner.

Robert DuBois
February 19, 2019

Parts of the Damselfly

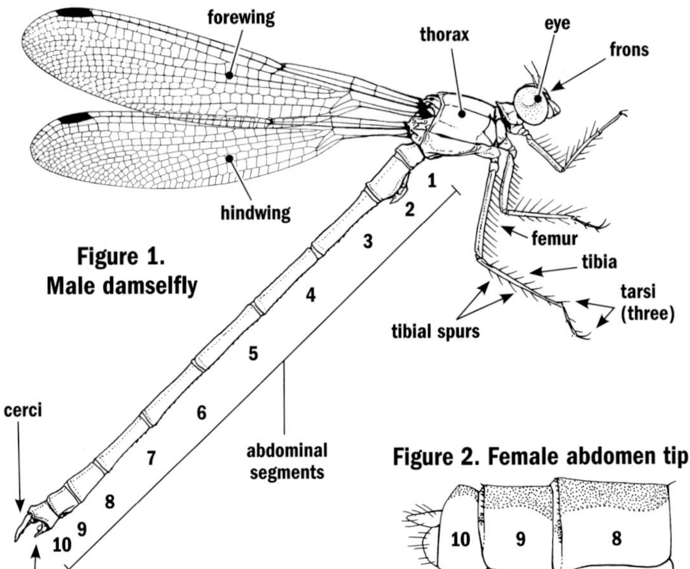

forewing
thorax
eye
frons
hindwing
1
2
3
4
5
6
7
8
9
10
femur
tibia
tarsi (three)
tibial spurs
cerci
abdominal segments
paraprocts

Figure 1. Male damselfly

Figure 2. Female abdomen tip

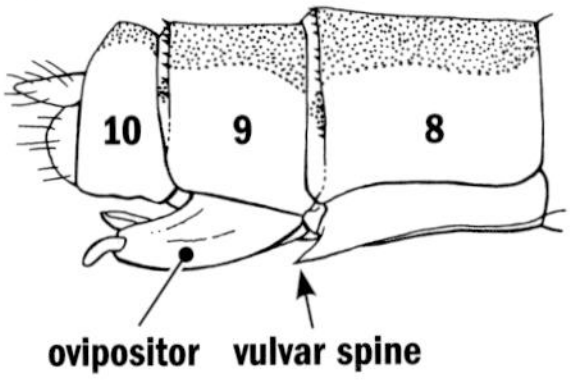

10
9
8
ovipositor
vulvar spine

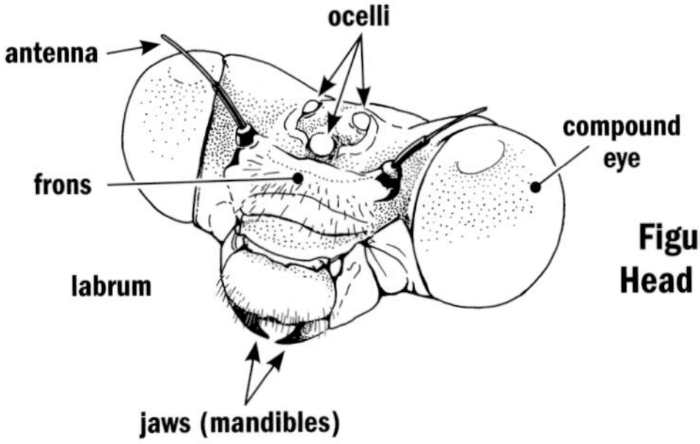

ocelli
antenna
compound eye
frons
labrum
jaws (mandibles)

Figure 3. Head detail

Figure 4. Wing detail

antenodal crossveins
nodus
Rs vein
stigma
arculus
M3 vein
M2 vein
M1 vein

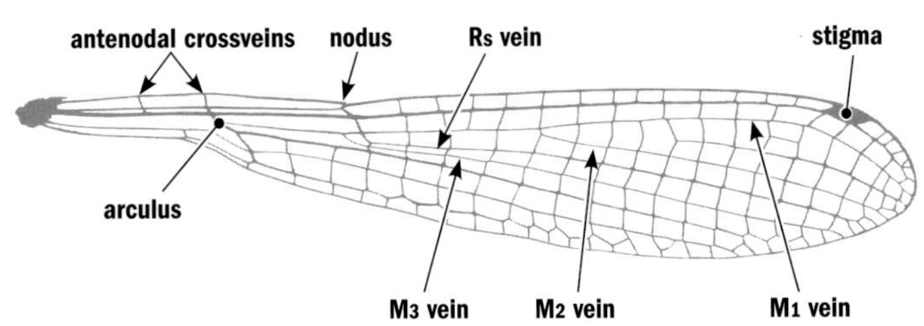

What is a Damselfly?

Remember when you were in high school and you learned about the biological classification system developed by Carl Linnaeus in 1758? Big groups were at the top of the chart that steadily got smaller toward the bottom: phylum and class, then order and family, then finally, the most specific categories of genus and species. Well, your science teacher probably explained to you why this terrifically successful system has remained in wide use for so long (with many modifications). One of the big reasons is that it gives us a framework for wrapping our minds around the staggering complexity and diversity of life. Otherwise, there would just be an enormous number of different critters out there and we would have no frame of reference for organizing and discussing this diversity. So, how do damselflies fit into this system?

Damselflies, like this Eastern Forktail, tend to hold their wings together up over their back.

Damselflies are joint-legged arthropods (phylum Arthropoda) within the class Insecta (insects), within the order Odonata, which means "toothed" or "toothed ones" referring to the rapacious jaws of damselflies and dragonflies. The common name "damselfly" comes from the French word demoiselle meaning "young mistress." I like to think that this refers to the slender, graceful appearance of damselflies. Within the Odonata are two Suborders of significance to us: the dragonflies (Anisoptera) and the damselflies (Zygoptera). Many people use the term odonate when referring to both dragonflies and damselflies.

Damselflies	**Dragonflies**
Eyes separated by at least own width	Eyes in contact with each other *(except clubtails)*
Hammer-headed, much wider than long	Bulbous-headed, nearly as long as wide
Long, slender build	Stout build
Quick, precise flight	Strong, sustained flight
Wings held over back when perched (or at 45 degree angle)	Wings held flat and perpendicular to body when perched.
Ovipositor present and functional	Ovipositor non-functional *(except in darners)*

To clarify how the two groups differ, let's consider what their scientific names mean. Anisoptera means "unlike wings," referring to the hindwings of dragonflies being larger than the forewings and differently shaped. In damselflies, the forewings and hindwings are similar in size and shape. This similarity is alluded to in the name Zygoptera, which roughly translates to "paired wings," from the Greek term zygon meaning "yoke." As you might guess from being in the same order, dragonflies and damselflies have much in common in terms of basic body plan and even many behaviors including reproduction and feeding. However, note the wing shape and the differences shown on the chart on page 1 to help you separate them.

Diversity

There are 20 families of damselflies worldwide containing about 2,700 named species. Many more species undoubtedly exist that have not yet been named, especially in the tropics. In North America north of Mexico we have five families with about 135 species. Here in Minnesota, Wisconsin and Michigan we have three families containing 51 species: four species of broad-winged damsels (Calopterygidae), eleven species of spreadwings (Lestidae), and 36 species of pond damsels (Coenagrionidae).

Damselfly Parts

Damselfly identification requires some knowledge about their anatomy. But before we go any further, let's get our directions straight: anterior = toward the head (front) end; posterior = toward the rear end; dorsal = top side, toward the back; ventral = bottom side, toward the underside; lateral = toward one of the sides (what would be seen in side view); basal = toward the base of a structure (end near the body); apical = toward the tip of a structure (end away from the body). The primary divisions of the damselfly's body are the head, thorax and abdomen, all covered with a hard (chitinous) exoskeleton.

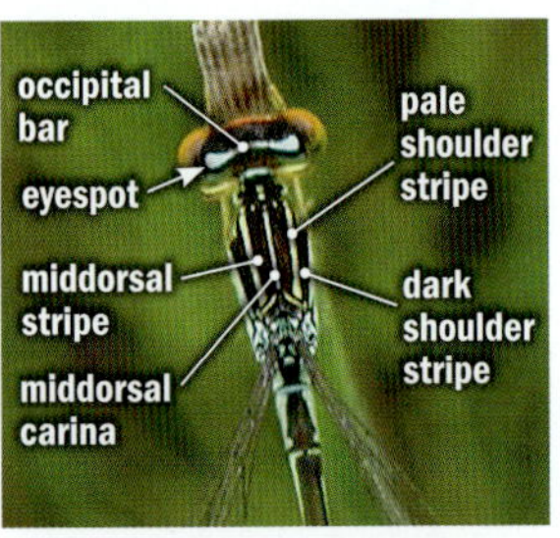

Knowing the head and thorax markings helps in identification.

Head

The head is much wider than long, with the large, compound eyes separated by a distance greater than the eye's width. This gives a damselfly a distinctly hammer-headed appearance. The huge compound eyes contain 5,000 to 10,000 individual light-sensitive units (ommatidia) and are well adapted for precise flight among vegetation and accurate perception of prey at short distances. They can detect the movement of light at twice the speed we can. It is now known that damselflies can see color, and they can also see ultraviolet light and the plane of polarized light. These features of the head and eye give odonates vision that is certainly among the best in the insect world. Three simple eyes called ocelli are also found on top of the head. Many species have pale eyespots, (postocular spots) on top of the head that are useful in identification. These eyespots differ in size among species and may be connected by a pale, transverse line called the occipital bar. The head is hollowed out at the rear and flexibly connected to the thorax so that it can rotate freely when scanning side-to-side and up and down in search of prey.

The "face" of the damselfly is made up of a number of chitinous plates separated by seams called sutures. From top to bottom the plates are known as the frons, postclypeus and anteclypeus (see Figure 3). Beneath the anteclypeus is the mouth, which is a complex but efficient "team" of parts that work together to get prey into the stomach quickly. The uppermost mouthpart (labrum) and the lowermost mouthpart (labium) hold the prey securely so the inner jaws can do the chewing. The inner jaws are comprised of a pair of stout, wedge-shaped mandibles, that tackle most of the heavy-duty chewing, and a pair of more slender maxillae. The antennae are small and bristle-like, composed of 6 or 7 segments. Although damselflies have no "smell center" like we do, they have recently been found to have tiny olfactory bulbs on their antennae that appear to help them track prey.

A male American Rubyspot stops and rests on the wings of an egg-laying jewelwing.

Thorax

The damselfly thorax is a unique, backward-slanting box that is made up of three sections. The small front section (prothorax) controls movement of the head and the front pair of legs. One of the signature features of damselflies is that the two rear thorax parts (mesothorax and metathorax) are fused together to form the box-like pterothorax (a.k.a. synthorax). This powerful "wheelhouse" contains the all of the flight musculature, to which the wings are directly attached, and it controls movement of the two hind pairs of legs as well. The structure of the thorax is skewed such that the legs are thrust forward and the wings backward, which aids in quick take-offs, the capture of aerial prey, and the gleaning of prey from surfaces while damselflies are in flight. The thorax is patterned in many of our species in ways that aid greatly in identification. A series of stripes form the foundation of these patterns in many genera. Our damselflies often have a relatively simple form of striping that consists, from top to bottom, of a mid-dorsal dark stripe along the very top of the thorax, a pale shoulder stripe known as the antehumeral stripe just below that, and a dark shoulder stripe called the humeral stripe below that. The mid-dorsal dark stripe may have a thin pale line along the mid-dorsal carina, which runs along the very top of the thorax. Some damselflies have no thoracic striping, others have the pale shoulder stripe reduced in the middle to a pair of dots, and still others may have the thoracic pattern partially or wholly obscured by **pruinescence** (see section on coloration or Glossary). Fieldmark arrows on the species pages will alert you to thoracic stripes that are useful for identifying those species.

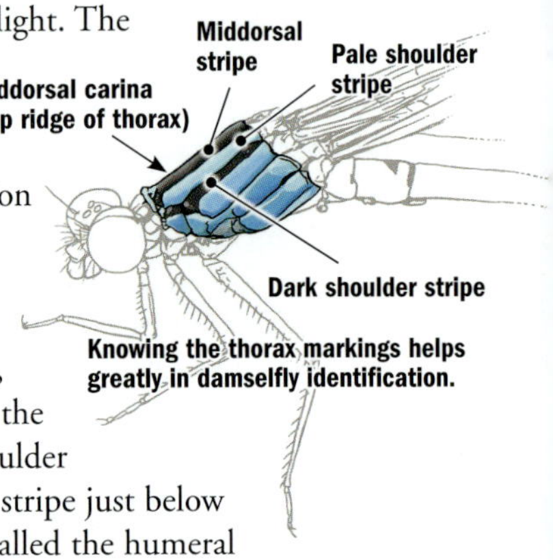

Knowing the thorax markings helps greatly in damselfly identification.

Wings

The wings of damselflies are flexible membranous structures, laced with tubes called veins that lend strength to the wing while at the same time allowing flexibility. When perched, the wings of damselflies are usually held together vertically above the body. However, the spreadwings (genus *Lestes*), rest with their wings partly spread at about a 45-degree angle to the body. The Taiga Bluet and Aurora Damsel also perch with their wings slightly spread. Damselflies never hold their wings perpendicular to the body (except for the Great Spreadwing, which comes close) or drooped downward, as some dragonflies do. Damselfly wings are generally narrowed (stalked) near the wing base, whereas dragonfly wings are not stalked, instead narrowing abruptly to the thorax.

Damselfly wings are directly attached to the muscles that move them, a condition called paleopterous (meaning ancient-winged). In contrast, most flying insects have the flight muscles directly attached only to the thorax and the wings move indirectly as a result of contraction of the thorax. This latter condition is called neopterous (meaning modern-winged) and permits more rapid wing movement.

This is one of the reasons why a house fly can flap its wings many more times per second than a damselfly can. Damselflies lack a hinge at the wing base and therefore cannot fold their wings flat over their backs.

One prominent feature of the wing is the pterostigma, or stigma for short, a darkened, blood-filled cell of uncertain aerodynamic function near the tip of each wing. What is certain is that the shape and color of the stigma is often useful in identification. All the wing veins of a damselfly—longitudinal veins, cross veins, and even some areas of vein intersection—have names and a thorough study of them takes time. But, learning just a few key veins and other wing parts (see pg. VI) will help immensely in identifying our northern damsels. Wings of our damselflies are usually clear, or hyaline, sometimes with a yellowish or amber "wash", but our broad-winged damsels have strikingly patterned wing markings that aid greatly in identification.

Legs

Like all insects, damselflies have six jointed legs. The long, upper leg section near the body is the femur. The next long section in the middle is the tibia. And the tarsi are comprised of three small sections making up the "foot." Rows of spine-like spurs along the femur and tibia are useful in capturing prey. The relative lengths of these spurs are useful for distinguishing the dancer genus (*Argia*) within the pond damsel family.

Abdomen

The abdomen of a damselfly is slender, cylindrical and always has ten segments, numbered 1 through 10, starting at the thorax. Patterns of color on these segments are often vitally important in identification, so it is well worth learning how to count up from the base, or back from the tip, which is typically easier, to find the segment in question. Segment 1 is usually very short, while segments 2 and 3 are longer and modified underneath in males to hold the secondary genitalia, a complex array of reproductive organs. The secondary genitalia are visible in side view on the underside of the abdomen base, and the presence of this "bulge" is an easy way to determine sex—females are smooth there. In all Odonata, sperm is produced in testes near the tip of the abdomen at segment 9, but the male must transfer free sperm to his own secondary genitalia prior to mating. This need for a male to transfer sperm from one end of his abdomen to the other, is unique in the insect world. With his "tool kit" of reproductive parts the male is not only able to deliver his own sperm to the female, but he can also first remove any sperm of other males left from previous matings.

At the tip of the male's abdomen are a set of claspers, or terminal appendages, which are comprised of the upper pair (cerci) and the lower pair (paraprocts). The shapes of these terminal appendages are crucially important in identification of most species of damselflies. The male uses his claspers to grasp the thorax of the female, usually at the mesostigmal plates of the synthorax, prior to mating. The tip of the female's abdomen also has pairs of cerci and paraprocts, but these are quite small and serve no known function in reproduction. The reproductive organs of the female consist of a genital aperture at the end of the segment 8, and a well-developed and fully functional ovipositor underneath segment 9. The ovipositor includes three sets of parts used to incise plant tissue and extrude one or several eggs into each cut. The shape and size of the ovipositor, and the presence or absence of a vulvar spine on the underside of segment 8, are useful aids in identification.

Telling Males from Females

Males

- "Bulge" visible on underside of abdominal segments 2 and 3 in side view.
- Abdomen relatively thin and slender
- "Claspers" at abdomen tip have upper and lower structures (magnification needed to see); their shapes very important in identification
- No ovipositor visible on underside of abdomen near tip
- Usually the more colorful sex when mature

Females

- No "bulge" on underside of segments 2 and 3 in side view
- Abdomen slightly stouter, bluntly tipped
- Only upper structure visible with magnification at abdomen tip, shape not useful in identification of our species
- Ovipositor visible as "bulge" on underside of abdomen near tip
- Often drab; usually less colorful than mature male

Biology & Behavior

The life cycle of a damselfly includes a lengthy aquatic stage as a nymph (also called larva or naiad), which is often a year or more in duration. Then, when seasonal and temperature cues are just right, the nymph emerges from the water and undergoes a quick and absolutely stunning transformation into a winged adult that will live for only a month or two at most. Despite their apparently gentle and serene demeanor, damselflies are ferociously predatory during both stages of life. In the long, slow nymph stage, the goal is pretty much just to grow. During the brief, but frenetic, adult stage, the primary goal is to reproduce. But to reproduce effectively, young adults must feed, gain weight and mature sexually, before they are ready to undergo the rigors of defending territories and competing for mates. Let's take a closer look at how this happens, first by peeking inside the bedroom door.

Mating

Damselflies mate only as mature adults. The required maturation period for newly emerged adults lasts from a day to several weeks, and usually takes place away from the breeding site, often in nearby clearings and forest edges depending on species. When ready to mate, males congregate at rendezvous sites at or near the breeding areas where they space themselves to await the arrival of females. Males of some of our species, including the jewelwings (genus *Calopteryx*) and some dancers (genus *Argia*), will actively defend territories from other males, as do many dragonflies. Most of our species are not aggressively territorial, though when two male forktails (genus *Ischnura*) interact, one may chase the other from the rendezvous site. Males of some of our species wait well inland where they attempt to intercept females before they reach the rendezvous sites. Hagen's Bluet is known to employ this sly tactic. Males of other species will patrol for females along pond edges without being territorial or engaging in aggressive behavior with other males. As a rule, females arrive at rendezvous or breeding sites after the males and only when ready to mate.

A male River Jewelwing courts a nearby female with his wing display.

Before mating (before or after tandem formation), the male must transfer free sperm from his testes on abdominal segment 9 to his secondary genitalia on abdominal segments 2 and 3. To do this he arches his abdomen until the appropriate segments meet. Fully loaded, he is now ready to initiate what is easily one of the most amazing spectacles in the insect world—formation of the mating wheel. The male uses his claspers to unceremoniously grab a female by the thorax. The pair is now "in tandem" and fly to a nearby perch. If she is ready to mate, she curls her abdomen forward so that the genital opening at the end of her abdomen engages the male's secondary genitalia. This produces a heart-shaped or circular configuration of their bodies commonly called being "in wheel" or "in copula." Copulation then proceeds for several minutes on average, but up to several hours for some forktails, most of this time being used by the male to remove any sperm of males that may have previously mated with that female. During the last few seconds of copulation, he transfers his own sperm into her sac-like bursa where it is stored until she actually lays her eggs.

A pair of Spotted Spreadwings in a "mating wheel," the heart-shaped position is typical of damselflies.

Mating between different species (hybridization) rarely occurs because of a variety of isolating mechanisms that block either tandem formation or copulation. Much has been made of the idea that a proper fit of the male's claspers with the female's thoracic plates is necessary for tandem formation, but of probably greater importance is that females have patterns of sensory areas on their thorax that match the shape of the male's claspers. This allows them to correctly identify males of their species and reject imposters.

Egg-laying and Guarding

The process of laying eggs is called oviposition. After mating, the male may leave immediately so that the female lays the eggs alone, or he may guard the female while she deposits the eggs. Two forms of guarding are used by damselflies in our region. The males of the broad-winged damsels release the female, but fly with her to the egg-laying site where he guards by hovering or perching nearby (non-contact guarding). In this strategy, he chases away competing males that attempt to mate with the egg-laying female, and he remains free to mate with other nearby females himself. By far the most common strategy for our species is for the male to remain in tandem while the female deposits her eggs (contact guarding). Either form of guarding enables the male to protect his genetic investment from being removed by competing males, and it allows the female to deposit eggs without being unduly harassed by other males. Females of most species mate repeatedly during the several weeks of their adult life. In a few species, notably our forktails, a single mating is apparently the rule because these females can store sperm for up to two weeks after insemination.

A single male Ebony Jewelwing watches over a bunch of females laying eggs. This is called non-contact guarding. Note his black wings that lack a white spot at the tip.

Female damselflies use their ovipositors to deposit eggs directly into the stems of rushes and sedges, into clumps of moss or into pieces of floating plant material. The female uses her blade-like ovipositor to "saw" punctures into the plant tissue (either above or below water) and extrude one or several eggs into each puncture. This is called endophytic oviposition. Females of species that lay eggs underwater may descend a foot or more and remain submerged for over one hour. Their supply of oxygen comes from a thin film of air that adheres to hairs on the body surface. Contact-guarding males may accompany females partially (usual case) or wholly underwater in a few species. Females of most pond damsels (family Coenagrionidae) rest horizontally on floating plant leaves when depositing eggs with the contact-guarding male standing rigidly upright as if "at attention," supported only by the grasp of his claspers on her thorax (known as the "sentinel" position, see photo pg. 19). By guarding in this way, the male is vulnerable to an attack from a predator, in which case his abdomen might break, leaving the lower part still attached to the female.

It is not uncommon for a male's abdomen to break off while in-tandem with a female, perhaps during the attack of a predator.

The Aquatic Nymph Stage

Damselflies spend the majority of their lives as slender, greenish or brownish, rather nondescript nymphs, crawling around underwater hunting their prey. They live in nearly every imaginable type of freshwater habitat including rivers, streams, lakes, ponds, vernal (temporary) pools, marshes, swamps, bogs, fens and spring seeps. The great majority of our species are found in still-water habitats, but with a handful occurring in rapid streams. Damselfly nymphs differ from dragonfly nymphs in several ways. They are more slender than dragonfly nymphs and unlike them, the head is wider than the thorax or abdomen. Three, slender-leaf-shape "gills" (caudal lamellae) at the tip of the abdomen help them breathe under water and swim gracefully in side-to-side fashion.

Damselfly nymphs are slender and long-bodied compared to the bulky, oval nymphs of their cousins the dragonflies.

The elongate eggs hatch rather quickly, usually from 12 days to five weeks after being laid, depending on air and water temperature, season and species. When an egg hatches, a tiny pronymph is released. Within just seconds or minutes, the exoskeleton of the pronymph is rent apart and the first true nymph stage, called an instar, bursts forth. Because the nymphal exoskeleton has little elasticity, the only way for the nymph to grow is to shed the old skin and replace it with a slightly larger one. As the nymphs feed and grow, species in our region progress through from ten to 17 molts before they are ready to transform into adults. Each time period between molts is called a stadium (plural stadia).

Tireless hunters, damselfly nymphs lurk on stems of standing aquatic plants or in beds of submersed vegetation, actively pursuing smaller insect prey or even tiny fish. To grab their prey, they are equipped with an elongate, hinged jaw (labium), which they are able to extend out with lightning speed, latching onto the prey with their raptorial labial hooks. Like so many things peculiar to the Odonata, this highly modified mouthpart seems to be unique in the insect world. Although capable of striking fear into the hearts of a vast array of small aquatic critters, damselfly nymphs do not bite or harm humans in any way.

While obtaining dinner is important to a damselfly nymph's survival, avoiding being dinner for some scary creature larger than itself is clearly a priority as well. Many animals dine on damselfly nymphs, including water birds, sunfishes, largemouth bass, and their own cousins, dragonfly nymphs. Studies have shown that ponds with sunfishes and largemouth bass (family Centrarchidae) tend to have different damselfly species than ponds without those fish species. It turns out that some species of damselfly nymphs try to escape from predators by swimming away, while others are content to crawl away and hide. Interestingly, the strategy of swimming away works well when escaping from insect predators, but not so well when the predators are fish.

Lyre-tipped Spreadwings depositing eggs into the stem of a rush as the males continue to guard their females.

The nymph stage typically lasts a year or less, but sometimes up to three years, depending on species. A general rule is that larger species, and species living in more northerly environs, take longer to develop. Most of our species overwinter as nymphs, except for spreadwings, the majority of which overwinter in the egg stage. The nymphs of our Eurasian bluets (Taiga Bluet, Subarctic Bluet, and Prairie Bluet) and perhaps a few species of American bluets are known to overwinter actually embedded in ice, where they are able to "enjoy" the winter without being bothered by predators. Some species of our spreadwings are adapted to live in temporary ponds; in these species the nymph stage may be completed in five months or less, before the ponds dry up.

Emergence

After months or years of clambering about underwater, the nymph is freed from the shackles of this ignoble existence in one grand moment of emancipation. Almost instantly it becomes one of the most graceful and elegant flying creatures under the sun. In the days before this transformation the nymph goes through many changes; it becomes more sluggish, stops feeding and the wing pads begin to swell. The mouthparts and other structures of the nymph become dysfunctional because the new adult (pharate adult) is now forming while imprisoned inside the nymphal exoskeleton. Just before emergence (technically the last ecdysis), the nymph rests at the water's surface with head and part of the thorax exposed and begins breathing through spiracles that have opened up on the sides of the thorax.

Unlike dragonflies, which often emerge at night or during the early morning, most damselflies emerge later in the morning or during midday. Nearly all our species climb a few inches above the water's surface, usually on aquatic plant stems, and emerge in a vertical (hanging) position. A few species of dancers climb onto logs or rocks and emerge in a horizontal (upright) position. To begin emergence, the nymph gains a secure grip on its chosen emergence support with its tarsal claws and rests for a few minutes while drying off. Then the thorax begins to bulge and splits longitudinally along its top. The top of the head also splits from side to side. These splits meet and then the thorax of the new adult, followed by its head, push out from this opening in the nymphal skin. Now free, the head and the thorax hang backwards for a few minutes while the new exoskeleton and the legs begin to harden. When the legs are strong enough, the new adult suddenly bends forward and grasps the emergence support or its own cast exoskeleton and pulls out its abdomen. The abdomen extends with air pressure from the gut and the wings unfurl and expand as insect blood (hemolymph) is pumped into them. Full expansion of the wings and abdomen takes up to an hour (less time with smaller species) before flight can be attempted.

A trio of Sweetflag Spreadwings (two males (above) and a female (below) temporarily left hanging). This is likely the top male's attempt to foil the other male's mating success.

The newly emerged damselfly, now called a teneral, is pale yellow, straw, or light green in color with no dark markings. Tenerals gain color quickly, but most species don't acquire full adult coloration until a few days after emergence (more quickly with forktails). Initially the flight is feeble and the wings glisten and shimmer in the sun. It is now very vulnerable to predation by birds, frogs, spiders, predatory insects, small mammals and even carnivorous plants like sundews. Some species emerge synchronously in huge numbers so that some of them make it through the gauntlet of predators. First flights are away from water as the tenerals scatter from the breeding site to feed, become sexually mature, and gain strength in nearby clearings, meadows and forest edges. Only when ready to mate do they return to the breeding site.

The cast skin, or exuvia (plural exuviae), remains on the emergence support, usually a plant stem, until it is carried away by wind, rain or high water. The exuvia retains the form of the mature nymph very well, and for that reason, can often be identified to species using nymph keys. Emergence of all individuals of a species usually occurs over a two- to three-week period, although some forktails (genus *Ischnura*) extend emergence over several months.

Coloration

One of the reasons why damselflies are so enjoyable to watch is that they come in such a striking variety of colors — beautiful blues, often with contrasting black markings, and iridescent greens are common. Other species in our region display various hues of yellow, orange, red, purple and brown. Colors and patterns play a huge role in identifying all our species. Patterns of coloration that contrast with the surrounding areas come in the form of stripes (narrow and elongate, running with the body), spots (roundish and usually small), and transverse bands (often called rings or annuli if they encircle a structure).

Learning to recognize our 51 species by colors and patterns is a bit more challenging than would initially appear. This is because males and females often do not look alike (sexual dichromatism). Also, tenerals of all species are much paler than mature adults, and lack dark markings. Male pond damsels are usually more brightly colored than their females. Identifying females to species can be quite tricky as there can be male-like forms and species that become grayish or bluish as they age (pruinescence). Pruinosity is a waxy substance that exudes from the cuticle. It is not limited to females; in spreadwings it is most pronounced in males. Male Powdered Dancers and female Lilypad Forktails become quite pruinose, to the point of becoming almost all white, as they age. The bright blues of some pond damsels can change to dull purple in response to cooler temperature, darkness, or while mating.

Parasitism

As you begin to observe damselflies closely you will soon notice that many have small orange or red egg-like bodies attached to the underside of their thorax or abdomen. These are not damselfly eggs but rather are water mite parasites that settle on the damselfly when in the nymph stage. When the damselfly nymph emerges, the larval mite is able to quickly crawl off the exuviae and transfer his boarding pass to the teneral damselfly. The mite now begins the parasitic stage, attaching to the new adult damselfly and sucking its body fluids for at least three or four weeks. It will then drop off back into the water to complete its life cycle. Parasites do not usually kill their hosts, but a heavy parasite load may hamper a damselfly's feeding and reproductive success. Pond damsels, along with the Skimmer family of dragonflies, are usually our most heavily parasitized odonates.

Feeding and Flight

Adult damselflies feed in two ways, the most common being the capture of flying insect prey on the wing using their forward-facing and spur-adorned legs kind of like a net. Some species also feed by gleaning small insect prey off the surfaces of foliage. Forktails are especially good at gleaning. Damselflies feed opportunistically on whatever prey species are available, showing little selectivity for specific prey items. A wide variety of flies are often the staple, with side dishes of caddisflies, small moths and other insects. Damselflies will certainly eat mosquitoes when they are available, both during the damselfly's nymph stage and when they are adults, but it is doubtful that they take enough to noticeably reduce populations. Once secured, the prey is carried to a nearby perch and rapidly consumed in rather large chunks. Contrary to the beliefs of some, damselflies neither sting nor bite hard enough to cause any serious pain to humans.

Though dainty in appearance, damselflies are capable of taking substantial prey items like this moth.

It is the remarkable and complex flying ability of damselflies that allows them to be very effective aerial predators. Although lacking the blazing speed of their larger cousins the dragonflies, damselflies excel at being quick, precise and extraordinarily maneuverable. They can fly in virtually any direction, even straight backwards. Their forewings and hind wings usually beat about 180 degrees out of phase, but they can also move all four wings synchronously or all four wings at different rates. Damselflies possess several flight advantages over other insects including an unusually high ratio of flight muscle to body mass and very low wing loading (the ratio of body mass to wing area). They are best suited to foraging low and slowly

A Mallard duckling happily pursues a bluet for lunch.

among vegetation, where their excellent vision allows them to home in on small prey at short distances. The habit of most damselflies to fly low, both amongst vegetation and over water, probably serves to reduce the risk of their being eaten by birds.

Orbweaver spiders and their large, sticky webs are a definite hazard to low-flying damselflies.

Migration

The kind of migration that odonates engage in consists of one-way movements from one place of residence (such as a breeding site) to another. Insects don't live long enough to move from one place of residence to another and back again, as do some birds, marine turtles and some fishes. Migration in odonates is imperfectly known and the movement patterns of some species are poorly understood. Some species seem to move just enough to find areas of less intense competition for resources and mates. A few species of dragonflies regularly engage in long-distance migrations. Damselflies are not generally recognized as being as overtly migratory as are some species of dragonflies. Nonetheless, several species are known to move around a lot, sometimes for considerable distances. Two species of damselflies in our region, Southern Spreadwing and Citrine Forktail, might be migratory, but much more needs to be learned about their life history. It is not known if either species consistently and successfully breeds here, or if they move into our region, either occasionally or regularly, from areas to our south.

Dispersal is an important facet of the ecology of many damselfly species. Dispersal just means flying to find new breeding sites rather than remaining at the old one. Probably all our species disperse at least to some extent, evidently to seek areas of less intense competition for food and mates. In our region, the Familiar Bluet is a particularly notable disperser that is very effective at finding and colonizing newly created or temporary habitats. Our populations have a bimodal flight period, and individuals in the later flight might be migratory. I have seen bogs and fens in northern Wisconsin that were densely populated with tenerals of this species in early September become totally devoid of any individuals just a week or two later.

Damselfly Observation

Why watch damselflies? Well, there are many compelling reasons, but among the best are that they are unique and often beautiful little creatures, fascinating in their behaviors, and unlike so many other animals, they are often easy to approach. Most species will offer the patient and careful observer a thorough look at all their sundry activities, which makes them wonderful subjects for in-depth nature study and macro-photography. Happily, most damselflies are active right when many of us like to be out and about—late mornings and afternoons on sunny, warm days. I like the way Kurt Mead put it in *Dragonflies of the North Woods,* "the best time for dragonfly observation usually occurs after a long, late breakfast, complete with an extra cup of coffee." Exactly so with damselflies as well! Indeed, they are card-carrying sun-worshipers—to the point that even a few dark clouds passing overhead will cause most of them to disappear so fast you'll wonder what happened. They don't actually disappear, but most will become inactive and perch in trees or bushes in the absence of considerable sunlight. However, a few of our species are crepuscular, meaning they fly until dusk or later.

Amber-winged Spreadwings in-tandem communally deposit eggs into a rush stem.

In the 14 years since *Damselflies of the North Woods* was written there have been two major developments that have greatly enriched the ways in which people can enjoy damselflies. First, digital cameras have improved in many ways and are more affordable than they used to be. The macro-photography capabilities of even relatively inexpensive point-and-shoot cameras and phones are now capable of producing images of surprisingly good quality. This advance has opened the door for many people to enjoy stalking and photographing damselflies. Second, there has been a great proliferation of Facebook groups dedicated to showcasing odonates and providing lots of help for beginners to improve their identification skills and increase their knowledge about behaviors and life history. Thousands nationwide are enjoying and benefiting from these groups. There are several good Facebook groups for damselfly enthusiasts to become involved with (see Damselfly Groups & Websites on pg. 150).

For the beginner, a good way to get started in damselflying is simply to pull on a pair of boots, grab this field guide, and your camera if you are so inclined, and head out to a nearby pond shoreline or other wetland on a sunny afternoon. Try to keep it simple and just enjoy the pleasures of discovery and exploration! Don't be afraid to hop right into their habitats. It is much more satisfying to immerse yourself in the damselfly's world than just watching their behavior from a distance. Perhaps the best tip I can give you is move slowly and pause often. You'll see much more that way. Observe the colors, sizes, behaviors and habitats of the damselflies you see flitting about you. Try to guess what they're doing and why. See if you can tell what family or genus they belong to. If you enjoy photography, you'll want to learn how to photograph your tiny subjects.

Rest stop. A small fly has landed on the convenient abdomen of a perched Amber-winged Spreadwing.

Habitats and Seasons

Damselflies can be found just about anywhere, particularly right after emergence when they scatter from breeding sites. But your best chance of finding lots of action is to go to favorite breeding areas. Well-vegetated pond and lake shorelines will reliably harbor a good number of species. Sedge Sprites, spreadwings, bluets and forktails are often found in these areas. Marshes and swamps will also hold many species of the same groups. Rivers and streams have different and fewer species than do still-water habitats. But don't miss out on flowing-water habitats because the species there are strikingly colored and absorbing to watch. Look for jewelwings, dancers, red damsels, a few species of bluets and Aurora Damsels along rivers and streams. Bogs and fens will often have sprites, some spreadwings and a few species of bluets.

Some species will be active each month from late May through early October in our region, however, the group of species present in the spring will be much different than those in the autumn. A few species (notably the forktails) have long flight periods that span much of the flight season, but most species have peaks of activity for only a month or two. Early-season flyers include the jewelwings, the Eurasian bluets, and some of the American bluets (genus *Enallagma*). Late spring and summer groups include the rubyspots, a few spreadwings, the dancers, most American bluets, some forktails and the sprites. Classic late-season flyers include most spreadwings and some forktails, with the Spotted Spreadwing invariably being the latest.

Gear and Equipment

As your interest in odonates blossoms, you are likely to find a few tools to be helpful. These probably will include a good pair of close-focusing binoculars, a camera with macro capabilities, a 10x hand lens (loupe), and an aerial net. The key words for successfully observing damsels with binoculars are "close focusing." Many companies make lightweight models that can focus down to six feet or less.

Close-up, or macro, photography has many advantages for the damselfly enthusiast including being an enjoyable pursuit in its own right, having use in learning identification, and in supplying a record of your trips. Digital cameras allow you to take large numbers of shots quickly, and to see what the results look like as you go. For purposes of identification it often helps to take multiple shots of a damselfly from several angles, so that the body part or parts needing to be seen to identify that species are in focus. Since the part(s) you need to be see will vary depending on the genus or species, you'll learn these finer points as you go. Much can be learned about how to get started with damselfly photography on the internet and from Odonata Facebook groups.

This male Tule Bluet is alertly defending his female as she lays eggs. This behavior is called contact-guarding and he is in the "sentinel" position

If you would like to identify most of the odonates you see to species, then sooner or later you'll want to get an aerial net and a 10x hand lens. The reason is simple—to identify many species, you must examine small body parts under magnification. Close-focusing binoculars will help, but won't always do the trick. Netting is the best way to get damselflies into your possession for a closer look. Although damselflies lack the lightning quick evasiveness of dragonflies, a bit of stealth is often still required. An underwear-ripping swing is not usually necessary—just a quick, deliberate stroke followed by a quick turn of the handle so the rim cuts off the insects' escape from the net bag. Most damselflies regularly perch on vegetation near the ground, so simply approach them quietly, and quickly drop your net over the top of them. Netted damselflies will usually fly upward toward the tip of an inverted net. Keeping the net rim low, reach your hand under the rim and use your thumb and forefinger to gently grasp the four wings of the damselfly together. In this way you can remove it from the net without damaging it. Now you can examine the specimen with a hand lens or prepare it to bring home to examine under a microscope. Very small species must be handled with considerable care to avoid injuring them, and tenerals should be handled only with discretion as they almost certainly will be damaged. Most mature damselflies can be released unharmed after being netted and handled carefully.

Look for a net with a deep bag, a rim with an opening of 12 to 15 inches in diameter, and a lightweight handle at least 2.5 feet long. Some nets come with extendable handles, allowing you to add segments to get any length you want. Others are collapsible for easy travel. Bear in mind that a longer, heavier net is harder to swing quickly, and swinging quickly is really the key. Children do best with a smaller net, with a 10- to 12-inch diameter opening. Some enthusiasts use much smaller nets with success, like the kid's bug nets stocked by larger department stores. Larger nets are more versatile, especially if you plan on netting dragonflies with the same net, and give you a larger margin for error when swinging. Smaller nets are easier to carry, may look less geeky, and might draw fewer spectators. Nets can be homemade or purchased from entomological equipment retailers. BioQuip Products stocks a variety of net sizes and styles, and they also sell replacement bags.

A vulnerable position. This male Stream Bluet is still contact-guarding his mate, though she has descended underwater to deposit eggs into an aquatic plant stem. I wonder if he's getting a bit nervous?

Using a 10x hand lens (loupe) takes a bit of getting used to, but if you'll make the effort, it will repay you in spades by opening up a whole new dimension of damselfly enjoyment. A loupe will allow you to identify many more damselflies than you could without one. The key is simply to hold the loupe up near your eye with one hand while moving the specimen held gently in your other hand until the damselfly and body part you need to see, comes into focus. That's it — move the specimen, not the loupe, to focus. Practice on a leaf or anything else that's not moving first.

As your interest in odonates deepens, you will find it to be advantageous to keep notes of your observations. It is best to record your observations immediately after an outing, before you begin to forget what you saw and where you saw it. If you take photographs of the species you see, you can often later associate the date or time stamp of the image with the species. Still, memory of some of the specific elements of your trip will fade over time. Taking notes will help you learn to identify and know the habits and flight periods of the species in your area more quickly. It will also allow you to contribute your observations to natural resource agencies and other groups that will make good use of your data. Taking notes provides enjoyment because it allows you to reflect on your outings, and provides a permanent record that you can look back on in the future. From May through October, I rarely go anywhere without my field notebook. Locations should be described as precisely as possible using latitude/longitude coordinates from a GPS unit, or relate the site to permanent physical entities like road crossings, bridges or boat landings. If you find a rare species, you can then report the site to the natural resource agency responsible for managing the area, and you will be able to describe the location to other people who might want to go to that location to observe that species. Other things you may want to record are time of day, weather conditions, behaviors

A pair of Variable Dancers laying eggs with Stream Bluets at a communal ovipositing site.

and stages of maturity of the species you see, evidence of breeding and habitat notes, including vegetation if you know something about it. If you keep up with your notes diligently, over time they will become increasingly valuable both to you and from a scientific perspective. People who work with dragonflies and damselflies professionally are quick to affirm that citizen enthusiasts have greatly increased scientific knowledge about species distributions, habitats and behaviors. Any notebook will do, but many people prefer Rite-in-the-Rain-type notebooks because they have waterproof paper. Always use pencil or water-proof ink because many inks will run when wet.

Proper dress is important to an enjoyable outing, including sunscreen and bug juice when appropriate. I like to wear a hat with a wide brim, lightweight field clothing and breathable waders that are scarcely heavier than slacks. I use knee-high, hip-high or chest-high waders depending on the water depth where I'm going. Some folks are perfectly happy just to wet-wade in tennis shoes—which is fine for them, just don't ask me to do it! Knee-high boots will suffice in many situations. A kayak, canoe or other small boat can get you into deep or muck-bottomed areas that would be impossible or unpleasant to wade, but be careful—swinging a net from a small craft can be tricky business.

Gaining access to good sites for damselflying can require a bit of planning. If you want access to private property, be sure to ask permission of the landowner first. When in doubt, ask—it's better to be safe than sorry. Depending on ownership, different public properties have different rules about public access and about the activities that are permissible on the property. Property access requirements may vary from state to state. To see detailed written guidance for property access in Wisconsin, go to the Wisconsin Odonata Survey website (http://atri-web.info/inventory/odonata). Property access requirements in other states and Canada may differ—be sure you do your homework.

Collecting

Although not for everyone, some damselfly enthusiasts choose to maintain a collection of these insects. Most scientists who study damselflies do so of necessity as well. The reasons for collecting specimens are many. Collected specimens, properly curated, are permanent and verifiable records of species distributions. It is important to document these distributions because they can change over time in response to losses of habitat, changes in climate, influences of invasive species and other factors. Management agencies charged with the responsibility of conserving and protecting natural resources for future generations need to know what species occur in different areas and what habitats are critical to their survival. Many state and provincial agencies now use citizen volunteers in surveys to help determine distributions and critical habitats of species (see page 150 for information about surveys in Minnesota, Wisconsin, Michigan and Ontario). Some collecting might be helpful when contributing to these surveys.

For some species, collection is necessary because careful examination under a microscope is the only way to identify them with certainty. Reference specimens are crucial for scientists who study identification and the phylogenetic relationships of species. Specimens are needed to properly determine relationships of closely related species and groups. Besides being essential to research, collecting specimens can be a source of pleasure and education. Responsible collecting poses no threat to damselfly populations, although restrictions on collecting are appropriate when populations are dangerously low, habitats are especially fragile or legal reasons apply.

In sum, collecting is a great way to learn, and it is a crucial tool in the protection and conservation of damselflies. If you are thinking about starting a collection, you should first read the Responsible Collecting Guidelines put out by the World Dragonfly Association and the Dragonfly Society of the Americas on their websites (see page 150). Then learn the proper techniques for collecting, preserving and storing specimens, also from these websites, or from a reliable odonatological text such as Paulson (2009) or Westfall and May (2006). When you are finished with it, a properly curated collection can be donated to a museum or university collection.

Equipment Sources

BioQuip Products: A wide selection of aerial nets (including the popular collapsible net), aquatic nets, Odonata envelopes, field guides, scopes and other entomological equipment.
www.bioquip.com 310.667.8800

Rose Entomology: A wide selection of entomological equipment including premium aerial nets.
www.roseentomology.com

Acorn Naturalists: A great source for aquatic dip nets, books and other naturalist and teaching resources.
www.acornnaturalists.com 800.422.8886

International Odonatological Research Institute: Best source for collecting envelopes. Also field guides and other books.
www.iodonata.net

How to use this Field Guide

Damselflies of Minnesota, Wisconsin & Michigan is designed to make field identification easier for you, the reader. Through the use of color photos, arrows pointing to field marks, size scales, phenograms and habitats, we have made a handy, compact and easy to use guide. Also, by limiting the damselflies to those found in one geographic area, we have eliminated the need to wade through several hundred species, many of which would never be found here. Included is every regularly occurring damselfly in Minnesota, Wisconsin and Michigan.

Order

Damselflies are organized by families and then broken down further into genera. Family name is listed at the bottom of the left page of each species spread while the genus is listed on the bottom of the right page. We placed closely related species next to each other. With experience in the field using this guide, you will gradually learn to identify damselflies to family, then genus and eventually, species.

Damselfly Names

Like other organisms, damselflies are given both common and scientific names. The common names are the English names most amateur naturalists use, while the scientific or Latin names tend to be the spoken word of odonatists. We capitalize species names but use lower case for groups (e.g. Stream Bluets are actually a type of bluet).

Photos

We chose to use photos of free-flying damselflies in their natural habitat. Males are always pictured on the left page of the spread and, with only a few exceptions, females are pictured on the right. The symbol used for males is ♂ and the female symbol is ♀. Photo credits are listed on page 150.

Fieldmark Arrows

Arrows point to diagnostic features in the photos that are referenced in the description text and marked with an arrow symbol (↑). These are characteristics that you should look for while in the field. Jotting down notes on size, color, thoracic stripes, abdomen markings and wing features will help you identify the damselfly when you have a chance to consult this book.

Abundance

A blue circle on each photo shows the relative abundance of each species: A=abundant, C=common, FC=fairly common, O=occasional, U=uncommon, R=rare and VR=very rare. Of course, even a "rare" species may be locally common if you happen to be in their preferred habitat at the right time.

Size Scale

Size is relative and often hard to judge in the field. Use the size-bars at the top of each species' main photo. The black bar indicates the actual body length of that species.

Phenograms

What is a phenogram? All damselflies live out their lives according to seasonal timing that is characteristic for that species. Our phenogram highlights in red the time when that species is active as a flying adult. In other words, look for that damselfly during the highlighted weeks/months. Of course, weather and latitude will cause some variation in emergence dates and flight period.

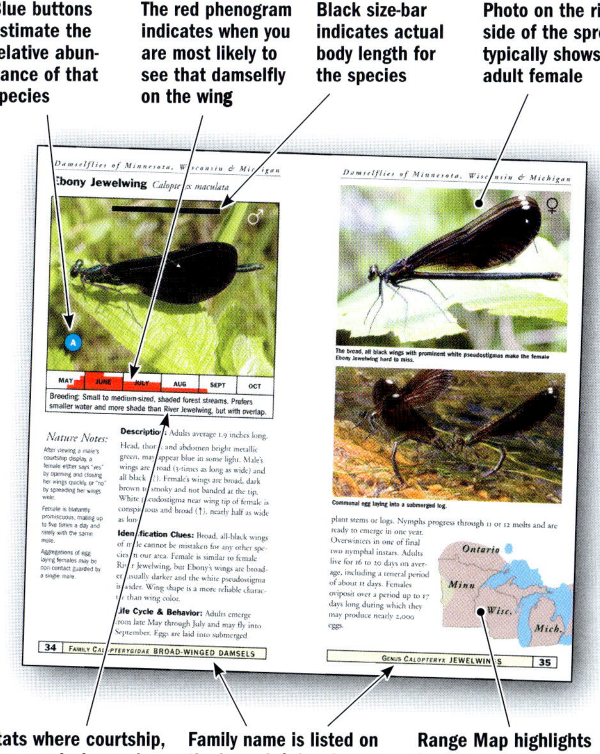

Habitat

Preferred habitat is found beneath the phenogram. Remember, this is where courtship, mating and egg laying takes place, and the nymph develops and emerges. Your best bet to find the species will probably be near such habitats—but not exclusively. Damselflies can range widely over many different habitats.

Nature Notes

Nature Notes are fascinating bits of natural history that bring one a more complete understanding of that species. Unique behavior, population trends and naming history are just some of the topics touched on.

Range Map

A map of Minnesota, Wisconsin and Michigan shows the range of that species in pink. This is the range as we know it from current population data but some damselfly populations are poorly understood.

Species Text

Description covers the best distinguishing characteristics first—whether it is the abdomen markings, shoulder stripes, wing color, veins or eyespots. If males and females are different (sexual dimorphism) then these differences are described. Measurements given are the average body length in inches.

Under **Identification Clues**, tips are given on how to separate that species from other similar damsels. Details of spreadwing male claspers (terminal appendages) and female ovipositors are shown side-by-side on pg. 43. Details of pond damsel male claspers are shown on pages 68-69. These features usually require a hand lens to see clearly.

Life Cycle & Behavior covers flight season, courtship, male guarding, egg laying technique and nymphal development.

Glossary

Check out the glossary for easy-to-understand meanings of some tricky terms.

Damselfly Checklist

Here you will find a checklist of all 51 species found in our three-state region of Minnesota, Wisconsin and Michigan. Check off the ones you see in your travels afield.

Enjoy *Damselflies of Minnesota, Wisconsin & Michigan.* Take it with on hikes. Stuff it in your canoe pack. Use it. But most importantly, have fun getting to know our fascinating northern damselflies.

Damselfly Identification

Many northern damselflies have readily observed markings or structural characteristics such that with the help of close-focusing binoculars or even the naked eye, you will learn to identify perched individuals without capturing them. However, some groups of species are so similar in appearance to each other that identification requires capturing them and examining small or awkwardly placed body parts closely in the hand, usually under magnification. Here are a couple of basic principles to remember. First, being able to determine the sex is almost always necessary (see pg. 6). Second, males are usually easier to identify than females and mature adults are easier to identify than tenerals. Of course, there are exceptions. For example, because of her outrageously long ovipositor, the female Sweetflag Spreadwing is easier to identify than the male. Bear in mind also that circumstances occur where even specialists may not attempt to identify every damselfly they see, especially the teneral females of some pond damsels. This is because the uncertainty of the result, and the risk of damaging the specimen, may make the effort less than worthwhile. Females of difficult genera can often be tentatively identified by identifying the males with which they are associating.

Three levels of identification are typically used with damselflies: in-the-field, in-the-hand and under-a-microscope. In-the-field refers to species that can be identified by field markings with the naked eye, with close-focusing binoculars, or through your camera in their natural state. In-the-hand refers to species that must be captured and examined more closely because the body parts or marks used for identification are too small to be seen at a distance, or must be viewed at a certain angle. For this level of identification, a 10x hand lens is normally used. Under-a-microscope refers to identifications that can only be reliably made with magnification of greater than 20x. This is particularly true of the mesostigmal plates on female pond damsels.

Damselflies of Minnesota, Wisconsin & Michigan will enable you to identify a great majority of our northern damselflies in-the-field or in-the-hand. This level of success has proved to be reasonably achievable for most nature enthusiasts. The under-a-microscope level is not addressed in this guide. As your interest in odonates deepens, you may at some point want to invest in a stereomicroscope, light source, and appropriate dichotomous keys. But you can cross that river when, and if, you come to it.

To identify a damselfly, first use the Family Keys and descriptions to determine the family of your subject (page 29). Then compare the body size, color patterns, terminal appendage (clasper) shapes, eye spot size, wing markings, and key aspects of wing venation, with the photos and illustrations in this guide. The broad-winged damsels can be identified in-the-field by their size, coloration, and wing markings. For spreadwings and pond damsels, capture some males with your net and examine their abdomen tips (terminal appendages) with a hand lens, comparing them with the illustrations in this guide. For male spreadwings, you need to look at the terminal appendages in dorsal (from the top) view; for male pond damsels, look at the terminal appendages in lateral (side) view. Also notice the patterns of stripes, spots and rings. First determine if the specimen is a spreadwing or a pond damsel using the Family Key on the opposite page. Spreadwings hold their wings partially open at rest, are usually larger than pond damsels, lack the kaleidoscope of colors of that group, and often are pruinose at the tip of the abdomen. Once you have your choice narrowed down to a group of species, use the identification clues given in the species accounts to rule out similar species. In this way, you can use this guide to identify mature males of virtually all damselflies in our three-state region and many of the females you will see.

I want to leave you with a final identification thought: it never hurts to net an odonate and examine its key structures closely even if you think you know what it is. Sometimes you will be surprised. Remember, color patterns within any given species can be vexingly variable, but structural parts, such as terminal appendages and ovipositors, generally are not. So, if you really need a positive identification, you should strongly consider capturing the specimen. As you get increasingly familiar with more and more species, there will be less need for netting. But netting and closely examining damselflies with a loupe is a great habit to get into when starting out, and it is a habit that experts stick with.

Family Key for adult Damselflies of MN, WI, MI

1a. Wings colored with black, brown or red, with no "stalk" at the base and with many antenodal cross veins (see drawing below); body metallic green or bronze; large species: BROAD-WINGED DAMSEL of FAMILY CALOPTERYGIDAE (pg.32)

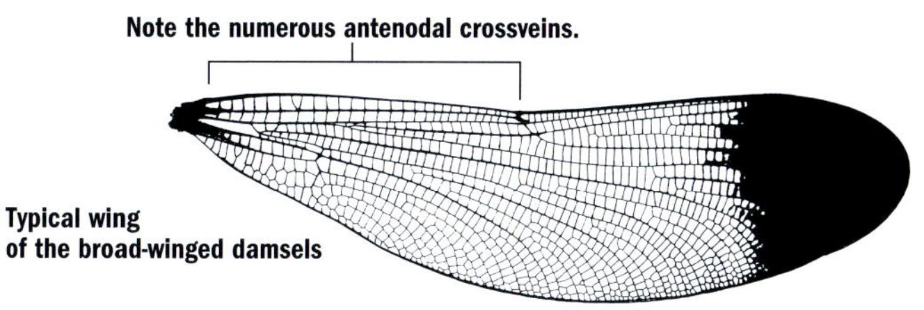

1b. Wings clear or lightly colored with amber, with a distinct "stalk" at the base and with just two antenodal cross veins; body may be any color; large and small species: go to 2

2a. Wings held partially open at rest; wing median vein splits closer to the arculus than the nodus (see drawing below); mostly larger species: SPREADWING of FAMILY LESTIDAE (pg. 40)

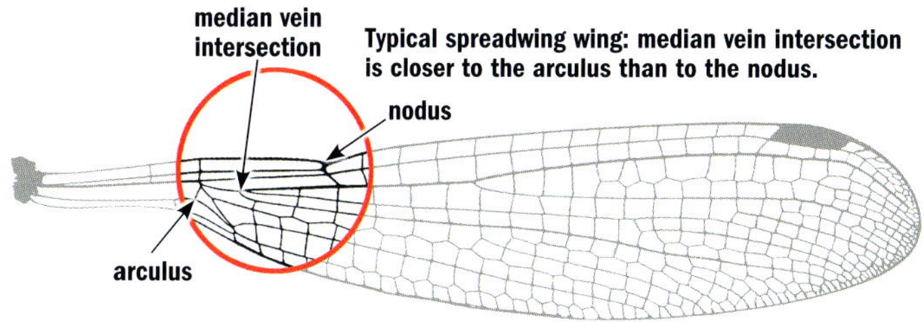

2b. Wings held together over the abdomen at rest (two exceptions); wing median vein splits closer to the nodus than the arculus (see drawing below); mostly smaller species: POND DAMSEL of FAMILY COENAGRIONIDAE (pg. 68)

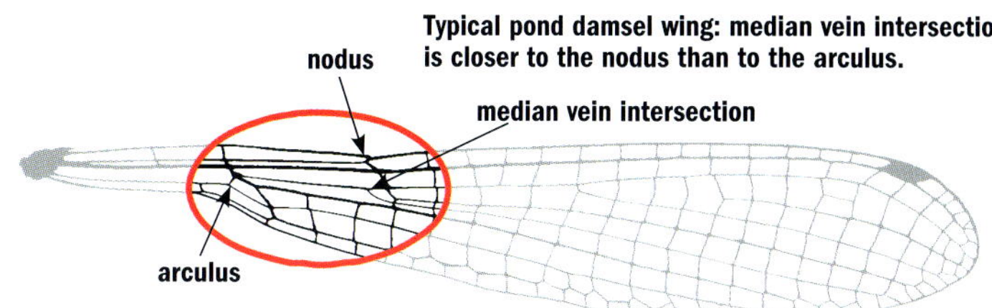

Broad-winged Damsels Family Calopterygidae

The broad-winged damsels are a small group of large, showy species that exclusively inhabit the flowing waters of rivers and streams. Some species prefer small to medium-sized streams, while others inhabit larger, slower rivers. They are readily separated from our other families of damselflies by easily seen characteristics of the wing, including color pattern, shape and wing venation. The family name, Calopterygidae, means "beautiful wing," referring to the striking patterns of wing coloration on all of our species. Wing shape in this family differs from other families in that the wings are broad and lack a distinctly narrowed "stalk" at the base. Among the many wing venation characteristics peculiar to the group is the presence of many antenodal crossveins on each wing, whereas other families only have two antenodal crossveins per wing.

All males of our broad-winged species engage in showy courtship displays and defend small territories along the stream's edge. None of the species venture far from water. At the breeding sites, females match males in numbers, unusual among the damselflies. Females choose whether to accept or reject a displaying male. Eggs are deposited into plant stems that are often well below the water's surface. Males do not remain in tandem during oviposition, but rather, guard in non-contact fashion by hovering or perching nearby.

The nymph stage lasts for at least a year and probably two or even three years for the River Jewelwing. The nymphs crawl about on submerged roots, plant stems and other forms of vegetation along the edges of flowing waters. Not surprisingly, the nymphs are large for damselflies and have broad wing pads. Nymphs in this family can be identified to species by characteristics of the labium, the length of antennal segments and presence or absence of abdominal spines.

About 180 species of broad-winged damsels are known worldwide (mostly tropical), with only eight species found in North America. Here in our region we have four species in two genera. Both sexes of all our species are identifiable in the field or in the hand by wing shape, color patterns, and for *Hetaerina* males, the shape of the paraprocts in side view.

The Jewelwings — Genus *Calopteryx*

The jewelwings are so named for the many species that have iridescent wings. Our two species have wings that are either partially or wholly black and the females show bright white patches (pseudostigmas) at the tips of their wings. Both sexes lack true stigmas. Their bodies are a brilliant, metallic green, with blue reflections in some light. Both species have legs that are long, slender and dark. Jewelwings are the only damselflies in our region with extensively black wings. Use the amount of black on the wings and wing shape to separate the two species, but this character is most useful for the males. For females, also use wing shape, and the relative widths of the white pseudostigmas near the wing tips to separate them.

Male jewelwings vigorously defend their "turf" along the edge of a stream, chasing away other males. In their territories they perform elaborate courtship displays, featuring special wing movements (see photo on pg. 7). Males will also alight on the surface of the water near the egg-laying site in what is called a dive display. This demonstrates to the female both the quality and location of the oviposition site. If a female is receptive to mate, she may allow the male to land on her closed wings, at which time he walks up along the wing's leading edge before taking her in tandem. To reject the male's display, the female simply spreads her wings and curves her abdomen up. The flight of jewelwings is usually described as bouncy and butterfly-like.

The Rubyspots — Genus *Hetaerina*

This genus gets its name from the unmistakable, blood red patches at the bases of the wings of males. The scientific name means "little companion," which Dennis Paulson and Sid Dunkle suggest may refer to the red armbands of Greek courtesans. The wings have venation patterns similar to jewelwings but are a bit narrower, usually have a small stigma and lack extensive areas of black (in our region, not true elsewhere). The bodies of both sexes are bronze to black. Males will defend their favorite perches from other males using a circle flight pattern. Female American Rubyspots will also defend territories; territorial behavior is unusual among female odonates. The main species in our region is the American Rubyspot, an unmistakable denizen of larger streams and rivers. However, keep your eyes open along the southern periphery of our region for the Smoky Rubyspot, which has a darker body, dark or smoky wing tips, and the male has an upturned tip of the paraproct.

River Jewelwing *Calopteryx aequabilis*

Breeding: Clean streams, medium-sized or larger, with open canopy. Also small to medium-sized rivers.

Nature Notes:

Male courts female with display of slow alternating wing beats. Male's courtship may include flinging himself onto the water's surface as if to show the female that the current speed in his territory is ideal.

Species name means "equal," perhaps referring to the nearly half black hindwing.

Description: Adults average 2 inches long. Head, thorax and abdomen bright metallic green, and may appear blue in some light. Male's wings clear at base, banded with black at the tip (↑). Hindwing's black band is larger (1/3 of wing) than forewing band (1/4 of wing). Dark wing bands may be larger on Canadian specimens. Female's wings are light brown to smoky and usually darker at the tip (↑). Her dark wing tips show less contrast with the rest of the wing than in the male. The white pseudostigma near wing tip of female is conspicuous but slender. Wings of both sexes 3 1/2- to 4-times longer than wide.

Identification Clues: Males are unmistakable; no other species in our region has a black patch at the wing tip. Female is similar to female Ebony Jewelwing, but River's wings are narrower, usually paler and the white

Although the female River Jewelwing's entire wing is often dark, the outer third is usually blacker than the rest.

pseudostigma is narrower (less than half as wide as long). Use wing shape more than wing color to separate females of the two species.

Life Cycle & Behavior: Adults emerge primarily in June and often fly into September. Eggs are laid into submerged plant stems. Female may crawl a foot or more below the water's surface to lay eggs and stay submerged for 40 minutes or more. Nymphs progress through 12 or 13 molts and are thought to require two or three years to develop.

Ebony Jewelwing *Calopteryx maculata*

Breeding: Small to medium-sized, shaded forest streams. Prefers smaller water and more shade than River Jewelwing, but with overlap.

Nature Notes:

After viewing a male's courtship display, a female either says "yes" by opening and closing her wings quickly, or "no" by spreading her wings wide.

Female is blatantly promiscuous, mating up to five times a day and rarely with the same male.

Aggregations of egg-laying females may be non-contact guarded by a single male.

Description: Adults average 1.9 inches long. Head, thorax and abdomen bright metallic green, may appear blue in some light. Male's wings are broad (3-times as long as wide) and all black (↑). Female's wings are broad, dark brown to smoky and not banded at the tip. White pseudostigma near wing tip of female is conspicuous and broad (↑), nearly half as wide as long.

Identification Clues: Broad, all-black wings of male cannot be mistaken for any other species in our area. Female is similar to female River Jewelwing, but Ebony's wings are broader, usually darker and the white pseudostigma is wider. Wing shape is a more reliable character than wing color.

Life Cycle & Behavior: Adults emerge from late May through July and may fly into September. Eggs are laid into submerged plant

The broad, all black wings with prominent white pseudostigmas make the female Ebony Jewelwing hard to miss.

Communal egg laying into a submerged log.

stems or logs. Nymphs progress through 11 or 12 molts and are ready to emerge in one year. Overwinters in one of final two nymphal instars. Adults live for 16 to 20 days on average, including a teneral period of about 11 days. Females oviposit over a period up to 17 days long during which they may produce nearly 2,000 eggs.

American Rubyspot *Hetaerina americana*

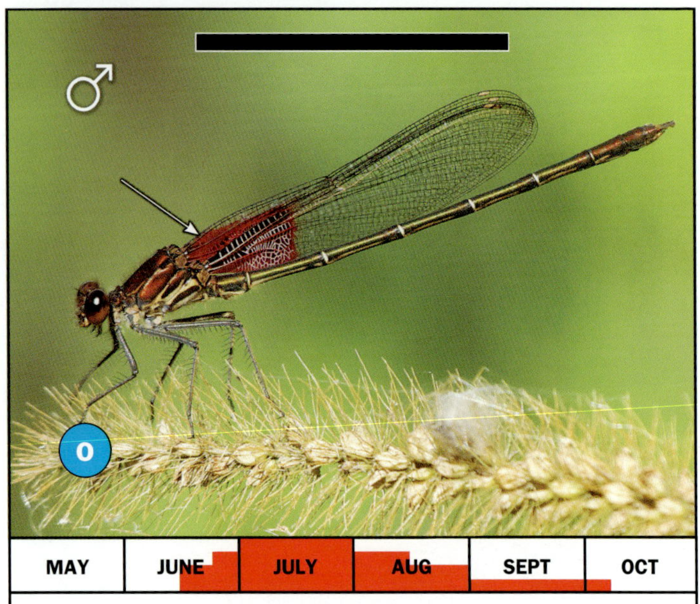

MAY	JUNE	JULY	AUG	SEPT	OCT

Breeding: Gently to rapidly flowing larger streams and rivers.

Nature Notes:

May congregate in densely packed groups. One Ohio odonatist described catching 75 with one sweep of his net!

Flits over shallow riffles where stream banks are well-vegetated. Does not wander far from stream.

Late season species that often flies through September.

May obelisk, pointing abdomen tip towards sun, at high temps.

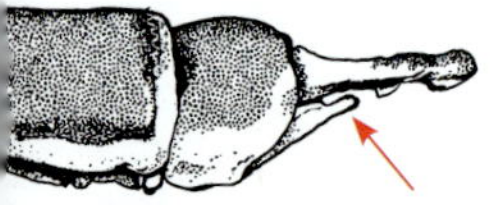

Male's paraprocts not upturned at tip

Description: Adults average 1.7 inches long. Male's head and thorax are red to reddish brown with thin pale lines along the sutures that become obscured with age. Wings are clear with a blood-red patch infused with white veins at the base (↑). Abdomen is metallic bronze to dark brown with pale, narrow rings on middle segments. Female's head and thorax variable—metallic reddish-brown, bronze or green with thin, pale lines along the sutures. Female's wings range from mostly clear to having yellowish to light brown wash, often with reddish or amber patch at base (↑) that often extends along leading edge (costal margin). Abdomen shades from metallic green to bronze, with pale, narrow rings on middle segments.

Identification Clues: Male can be confused only with Smoky Rubyspot, but blood-red patches at wing bases average larger, thorax and

Females usually have a metallic green thorax (top photo) but some are metallic red like the one in the bottom photo.

abdomen not as dark, lacks any trace of black at the wing tips, and paraprocts are not upturned at tips (see illustration on opposite page). Female much like Smoky Rubyspot, but lighter on sides and venter of abdomen, lower legs are pale, and dark marks on ovipositor are scant or lacking.

Life Cycle & Behavior: Adults emerge from late June into August, and may fly into October. Eggs are laid underwater, usually a few inches under but sometimes up to 60 cm deep, into plant stems or partially submerged, rotted wood. Female may stay submerged for nearly an hour. Nymphs likely overwinter in the 2nd to 4th instars and are ready to emerge in one year. Males use circle flight pattern to defend favored perches from other males.

Smoky Rubyspot *Hetaerina titia*

MAY	JUNE	JULY	AUG	SEPT	OCT

Breeding: Gently flowing larger streams and rivers in forested landscapes.

Nature Notes:

Prefers slower, deeper river sections lacking riffles and shadier areas than American Rubyspot.

Territorial males use circle flight pattern and flick wings to ward off other males.

Highly variable wing colors have been experimentally shown not to change over time.

Scientific species name means having a reddish-brown color, which may refer to the male's wings.

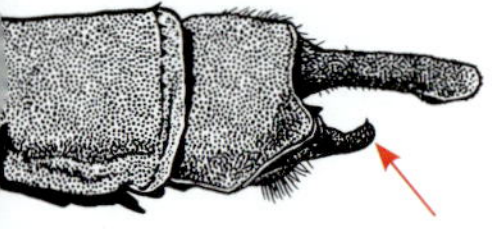

Male's paraprocts are upturned at tip

Description: Adults average 1.7 inches long. Male's head and thorax are dark brown to black, with dull green or red reflections and with very fine pale lines along the sutures that become obscured with age. Wings variably marked ranging from black to mostly clear; clearer winged specimens will show blood-red patch at the base of forewing, brown patch at base of hindwing (↑), and dark smudge at wing tips (↑). Abdomen is dark brown to black with pale, very narrow rings sometimes visible on middle segments. Female's head, thorax, and abdomen similar to male's, but not always as dark and thorax often shows some green. Her ovipositor is mostly or all dark. Female's wings similarly variable, ranging from nearly clear to having dark brown wash throughout, often with brown patch at base (↑). It is not known if the great variation in wing marking of both sexes are due to season or location.

Some females have dark tinted wings (inset).

Identification Clues: Dark-winged specimens could be confused with jewelwings, but wings are narrower and some thoracic markings usually visible. At the southern periphery of our three-state region, clearer-winged specimens seem to be the rule. Clearer-winged males can be confused only with American Rubyspot, but red patch at hindwing base usually smaller and darker, wing tips show some back, thorax and abdomen are darker with pale lines along sutures less apparent, and paraprocts are upturned at tips (compare with illustration on opposite page). Female often much like American Rubyspot, but abdomen is darker on sides and venter, lower legs are dark, and ovipositor is mostly or all dark. When in doubt about female identity, look at associated males.

Life Cycle & Behavior: Adults emerge from late June into August, and may fly into September. Females oviposit under water, staying submerged for up to 2 hours, with male guarding above in non-contact fashion. Little is known about the life history of this species at northern periphery of its range. Nymphs likely overwinter in a later instar and are ready to emerge in one year.

Spreadwings, Family Lestidae

The spreadwings differ from all other damselflies in our region in that they perch with their wings held partially open, at about a 45-degree angle to the body (hence their common name). The only other adult damselflies in our area that perch with wings slightly spread are the Taiga Bluet and Aurora Damsel, which are easy to separate from the spreadwings by their color patterns, smaller sizes, and wing venation. A word of caution — don't use the open-winged position to identify tenerals, because they refuse to play by the rules. Some teneral pond damsels hold their wings spread when perched, like spreadwings and newly emerged spreadwings hold their wings together over their backs like pond damsels.

Spreadwings are medium to large in size and are usually found at still-water habitats including temporary ponds. They like to perch on vertical plant stems holding their bodies downward, usually at an oblique angle. Spreadwings usually are not brightly colored, although the bodies of some are a wonderful iridescent green or bronze. Others are gray, brown or black, and most have various amounts of pruinosity. These cryptic tones blend in well with the rushes, reeds and sedges of their preferred haunts. Mature males often have bright blue eyes and the tips of the abdomens of most species appear white or grayish-white because of pruinosity. Males tend to become increasingly pruinose with age, often obscuring the underlying colors of the thorax. The wings of spreadwings are narrow, distinctly stalked at the base (called petiolate), not heavily veined and generally clear. The exceptions are the Great Spreadwing and Amber-winged Spreadwing, both of which have an amber-colored or smoky wash on their wings. The wing venation of the group is unique because the position on the wing where the median vein splits sets them apart from pond damsels (see pg. 29), as does their long stigmas.

About 152 species of Lestidae are known worldwide, of which 19 are found in North America north of Mexico. Eleven species, ten in the genus *Lestes* and one in the genus *Archilestes*, are treated in this guide.

The Stream Spreadwings — Genus ***Archilestes***

These are very large damselflies, the largest in North America, and they look much larger than any northern pond spreadwing or pond damsel. Prominent yellow stripes on the sides of the thorax set our single species apart from all other members of the family. These stripes are easily visible, not becoming obscured by pruinosity, which develops only on the abdomen tip of the male. Males defend small territories. They are primarily stream species but are often found at ponds as well. Females are unusual in laying their eggs into woody stems that may be far above water.

The Pond Spreadwings — Genus *Lestes*

The pond spreadwings are hard to confuse with any other group. No other medium-sized damselflies perch on vertical stems with abdomen lowered and wings held apart. Also, the blue eyes and grayish-white (pruinose) abdomen tips of males of most species are unmistakable. When taking a close look at a member of this genus, notice that the spurs on the lower legs are unusually long — much longer than the spaces between them. Adults are usually found along cattail-, rush-, sedge- and reed-bordered shores of marshes, temporary and permanent ponds, and small, sheltered lakes. A few species are infrequently found on slow sections of rivers and streams. They flit from stem to stem along the shore in short, low, flights, feeding primarily on small flies that have emerged from the same areas. Territoriality and courtship behavior are absent or poorly developed in this genus.

The pond spreadwing life cycle has some interesting aspects stemming from how the various species portion out the habitat they use depending on the permanence of the water body and kinds of predators that are present. Most of our species do best in waters with few or no fish, where the top predators are usually dragonfly nymphs in the darner family. But such waters are often temporary for some portion of the year (or they would have fish). These species deal with temporary waters by laying their eggs late in the season and spending the winter in the egg stage, which no other damselfly genus in our region does. This egg-wintering strategy allows them to utilize temporary ponds and vernal pools that dry up during the summer or fall, because the eggs do not need to be wet all the time. Female pond spreadwings of the egg-wintering variety oviposit, in tandem with the male, into the stems of cattails, bulrushes or spikerushes often a foot or more above the water (except the Northern Spreadwing does so submerged). The eggs immediately take some of the initial developing steps, but before hatching occurs they slip into a state of embryonic diapause (rest). In this state, the drying-resistant eggs spend the winter inside plant stems where they withstand temperatures as cold as it gets in our region (with help from an insulating cover of snow). Then, after the snows melt and water levels rise in the spring, the dead plant stems are wetted, providing the moisture needed along with the right increasing temperature and day-length cues to complete the hatching process. The nymph stage is short for these egg-diapausing species — just two or three months in spring and early summer — because the fast-growing nymphs need to complete their development of about ten instars and emerge before the temporary ponds they live in dry up. The Northern, Southern, Sweetflag, Slender, Lyre-tipped, and Spotted spreadwings typically fall into this category, although some of them are sometimes found in permanent ponds and streams as well, particularly if they can find places where fish densities are low. The

Emerald Spreadwing does best in temporary (vernal) pools that lack both fish and dragonfly predators. Even though these pools dry out by early summer, the nymphs have such explosively fast growth rates that they can complete their development and emerge before then.

Contrastingly, Swamp and Elegant spreadwings, like many other damselflies, do just fine in permanent waters with fish predators. They lay their eggs in spring or early summer into plant stems near or above the water's surface and the egg hatching process soon starts. Permanent waters help these species because the eggs hatch long before winter and the partly grown nymphs must spend the winter underwater (which is typical of the great majority of odonates). Species that overwinter as nymphs have a long nymphal life of at least nine months and up to 17 instars.

Pond spreadwing nymphs are physically unique as well. They are long, slender, and agile, with long legs and large, leaf-like caudal lamellae. Flip one on its back and you will see an amazingly long labium (face mask) that is strongly narrowed at its base. No other damselfly in our region has a labium quite like it. The nymphs cling to submerged vegetation where they patiently stalk their prey.

Males of all species can be identified in the hand by the definitive shapes of their terminal appendages (claspers) seen in dorsal view, in conjunction with other characters. Some species (both sexes) can be identified in the field because of their unique body shape or size, body coloration, wing coloration, or elongate ovipositor. Note however, that pruinosity increases with age for both males and females, which can complicate species identification. Tenerals and immatures are particularly problematic to identify because they are poorly marked and lack the colors they will have as adults. Further, one group of similar species creates some serious identification challenges; this group includes the Northern, Southern, Sweetflag, and Slender spreadwings. Males in this group are generally identifiable in the hand by differences in the shapes of their terminal appendages, except that Southern and Sweetflag can be difficult to separate. Females in this group, with a single exception, can be identified with certainty only under a microscope, primarily by measuring subtle differences in the lengths of some abdominal segments relative to the length of the ovipositor, and by colors of some small body parts. However, the female Sweetflag is readily identifiable because she has an extraordinarily long ovipositor. For each species then, either the male or the female can be identified, so identifying one sex reinforces the identity of the other. Therefore, look for tandem pairs when dealing with this challenging group. The name *Lestes* means "robber" or "pirate," probably referring to the predatory habits of adults or nymphs.

Top view of male claspers

Side view of female abdomen tip

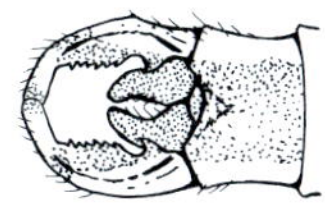

Spotted Spreadwing
Lestes congener
pg. 46

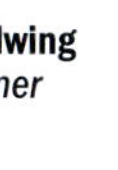

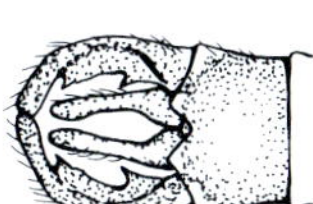

Northern Spreadwing
Lestes disjunctus
pg. 48

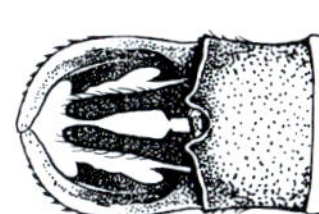

Southern Spreadwing
Lestes australis
pg. 50

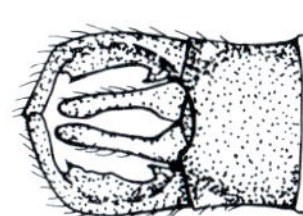

Sweetflag Spreadwing
Lestes forcipatus
pg. 52

Lyre-tipped Spreadwing
L. unguiculatus
pg. 54

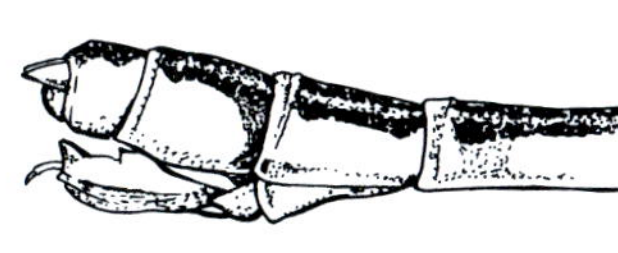

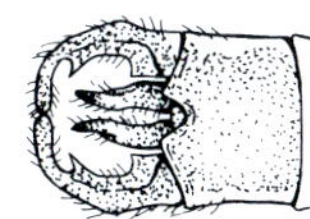

Slender Spreadwing
L. rectangularis
pg. 56

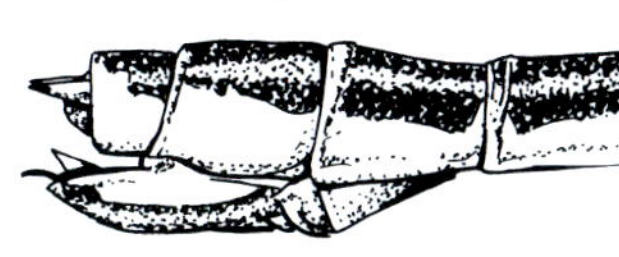

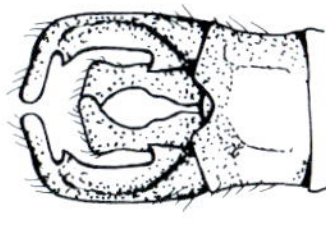

Emerald Spreadwing
Lestes dryas
pg. 58

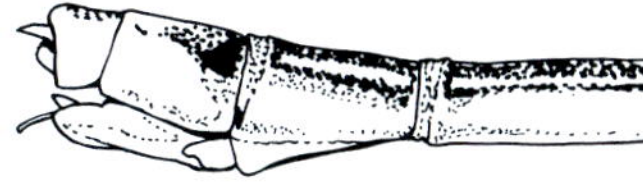

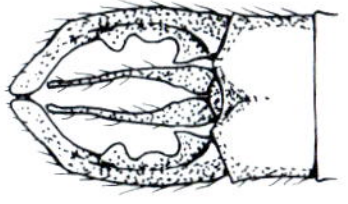

Swamp Spreadwing
Lestes vigilax
pg. 60

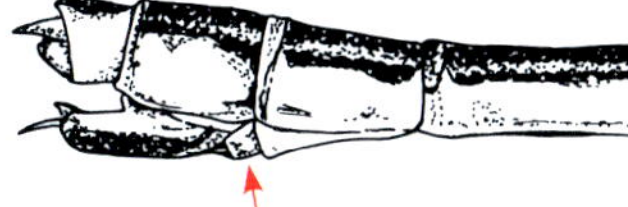

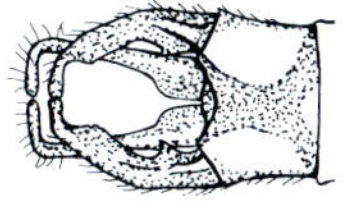

Elegant Spreadwing
Lestes inaequalis
pg. 62

Note angled margin of basal plate

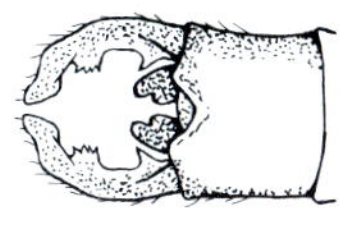

Amber-winged Spreadwing
Lestes eurinus
pg. 64

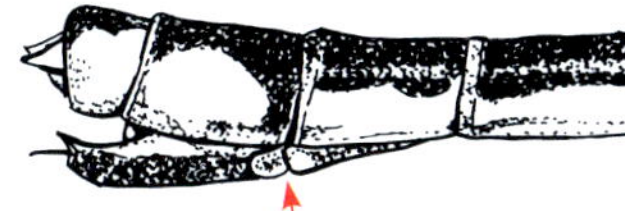

Note rounded margin of basal plate

Great Spreadwing *Archilestes grandis*

MAY	JUNE	JULY	AUG	SEPT	OCT

Breeding: Slow streams, stream backwaters, ditches, and some ponds in forested areas.

Nature Notes:

Largest North American damselfly.

Has dramatically expanded its range across the nation from the southwestern US since the 1920s.

Female has been observed to lay eggs in a tree, 44 feet above water!

The robust nymphs swim actively like tiny fish.

Description: Adults average 2.3 inches long. Male has blue eyes, powder blue labrum, and dark brown thorax with thin, metallic green shoulder stripes (↑) and two bold, yellow stripes on side of thorax (↑). Thorax shows light coat of pruinosity as male matures. Male's abdomen dark brown or black with gray pruinose abdomen tip on segments 9 & 10. Wings clear or with smoky tinge. Male's claspers feature very short paraprocts, not easily visible is dorsal view. Female marked much like male but lacks pruinosity. Her eyes may be blue or brown and abdomen has pronounced bulb-like tip. Ventral margin of her ovipositor has large, coarse "teeth."

Identification Clues: Both sexes are unmistakable because of large size, prominent yellow stripes on side of thorax, and widely

Bold yellow stripe on each side of her thorax instantly identify this female.

outspread wings. Looks enormous compared to other damselflies!

Life Cycle & Behavior: A late season species that emerges during summer and flies into September or later. Female lays eggs in tandem with male into herbaceous or woody stems or leaf petioles often 3 feet or more above water. One pair oviposited in tandem for an hour then the male left and the female continued alone for another 30 minutes. Egg overwintering is the rule in the north, but nymphs may overwinter in the south. Nymphs quickly complete development and mature in one growing season. Prefers fishless waters where nymphs swim actively in absence of fish predation. Species is known to tolerate poor water quality, which may also correlate with reduced numbers of predators and competitors at some sites.

Spotted Spreadwing *Lestes congener*

Breeding: Permanent and temporary still-water habitats: marshy and bog-bordered ponds, swamps, marshes and slow streams.

Nature Notes:

Our latest damselfly, often seen into mid October or later.

Common and widespread in our region.

Species name means "of the same kind," possibly to include them with others of the same genus.

In forested landscape will roosts high in trees, to a height of 10 feet or more.

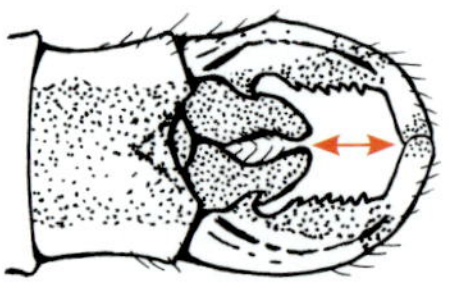

male's claspers; cerci about twice as long as paraprocts

Description: Adults average 1.4 inches long. A small, dark and stocky pond spreadwing. Thorax and abdomen of male is dark above changing abruptly to pale whitish gray sides and a pair of dark spots on lower edge of thorax (↑). Thorax has very thin, pale shoulder stripes. Pruinosity on the thorax becomes more extensive with age. Abdomen is dark brown above, lighter on sides with last two segments grayish-white pruinose. Male cerci about twice as long as short paraprocts (see illustration this page). Wings clear; his eyes blue above. Female is similar to male but with a thicker abdomen and brown eyes. Female ovipositor is short, not extending beyond tip of abdomen.

Identification Clues: Both sexes of this small, dark, late-season species are easy to identify by the presence of a pair of dark spots on the lower edge of the thorax. Male Sweetflag Spreadwing

Both the male and female Spotted Spreadwings show a distinctive pair of spots along the lower edge of the thorax. The Sweetflag Spreadwing may have a single spot here.

usually has one dark spot there, but is larger and has much longer paraprocts. Dark look of male thorax on top contrasts sharply with very light sides, and his blue eyes often stand out as well.

Life Cycle & Behavior: Adults begin to emerge in July and peak abundance is reached in September. Some fly well into October, even November some years. Pairs form away from water then fly close to the egg-laying areas. Eggs are laid into dry bulrush, spikerush, or cattail stems above water level. Overwinters in egg stage, but in an earlier state of embryonic development than other spreadwings. Eggs survive air temperatures at least as low as -18° F, colder when snow covered. Eggs hatch in late spring, at which time the pronymph emerges from the cut in the stem and quickly molts into the second instar. Nymphal development through ten instars takes about 50 days. Newly emerged adults require three weeks to mature.

Northern Spreadwing *Lestes disjunctus*

MAY	JUNE	JULY	AUG	SEPT	OCT

Breeding: Well-vegetated, marshy or boggy ponds that may be permanent or temporary. Also bogs, fens and slow streams.

Nature Notes:

One of the most abundant and widespread spreadwings in our region.

It was formerly called the Common Spreadwing, at which time it was combined with Southern Spreadwing.

Only spreadwing in our region that often oviposits underwater.

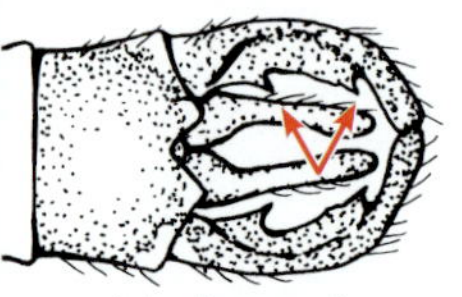

male's claspers; two equal teeth on cerci

Description: Adults average 1.5 inches long. A mid-sized, dark pond spreadwing. Thorax and abdomen of male is dark above with pale yellow or gray sides. Thorax has pale greenish-blue shoulder stripes (↑) and a thin pale line along top. Dark color on thorax becomes more extensive with age and gains some pruinosity. Abdomen is dark with metallic green reflections and the last two segments are grayish-white pruinose. Wings clear. Female is similar, with sides of thorax yellow to light tan and her abdomen is stout and dark above. Rear of female's head is mostly dark, with pale areas only around neck. Female ovipositor is short, not extending beyond tip of abdomen, and is mostly pale, sometimes with dark smudge in middle (↑).

Identification Clues: Two equal-sized and shaped "teeth" on inner margins of male cerci,

Bottom half of female's ovipositor less likely to be all black than on other similar species.

seen in dorsal view with a hand lens are diagnostic. Sweetflag and Southern spreadwing males are similar, but have the two teeth on cercus further apart and end (distal) tooth less pointed. They usually average slightly larger in size as well. Further, Northern usually entirely pruinose on top of segment 2 when mature, Sweetflag not so, showing shiny area at apical third. Also, apical notch on segment 10 of Northern looks narrower and more pinched than wider notch on Sweetflag. Female Northern Spreadwing is similar to females of several other species, differing as follows: Lyre-tipped has more yellow at rear of head that extends laterally reaching eyes; Sweetflag has much longer ovipositor; Southern has darker bottom half of ovipositor (Northern mostly pale there); and Slender has pale areas on tarsi (Northern all black there). Positive identification of females of these species is based on subtle differences in lengths of some abdominal segments relative to the length of the ovipositor, measured under a microscope. Use identity of males with which they are associating to tentatively identify questionable females.

Life Cycle & Behavior: Adults emerge in late June and July and fly into September. Eggs are laid in tandem, usually into the green stems of bulrushes, often below water level. Eggs are cold-tolerant, but need insulating snow cover; hatch in spring when water temperatures reach about 50 degrees F. Nymphal development through ten instars takes about 60 days and maturation requires 16 to 18 days after emergence.

Southern Spreadwing *Lestes australis*

Breeding: Wide variety of well-vegetated or boggy ponds, lakes, and marshes; occasionally in slow streams.

Nature Notes:

Consistently the earliest spreadwing seen in spring in our region.

Usually seen here in low numbers; species could be migratory or stray from areas to the south.

When observing this and similar species, look for tandem pairs because as Ed Lam noted, identification of one sex usually reinforces the identity of the other.

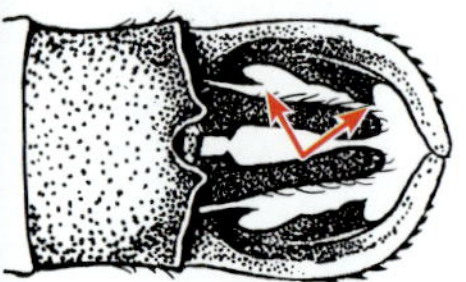

male's claspers; two unequal teeth on cerci

Description: Adults average 1.6 inches long. A mid-sized, dark, early season spreadwing. Thorax and abdomen of male dark above with pale yellowish sides becoming grayish-white with pruinosity. Pale shoulder stripes tan, light green, or blue, probably depending on age. Abdomen has dark ventro-lateral spots usually showing strong dark/light contrast in side view. Last two abdominal segments and part of S8 are grayish-white pruinose (↑). Wings clear. Female colored like male, but not becoming pruinose, with sides of thorax yellowish, and her abdomen stouter. Rear of female's head is dark with pale area only around neck. Female's ovipositor is short, not extending beyond abdomen tip, with lower half black (↑).

Identification Clues: Separate male from male Northern Spreadwing by Southern's two unequal teeth on cerci, end tooth is smaller and blunter than basal tooth; also, teeth are further

Lower half of female's ovipositor is black.

apart (see illustration on opposite page). Male very similar to male Sweetflag Spreadwing with identical teeth on cerci, but Southern typically has paraprocts slightly incurved in dorsal view, unlike Sweetflag. Focusing on tandem pairs extremely helpful in separating these species as female Sweetflag has unmistakably long ovipositor while that of Southern is of usual shorter length. Early flight period of Southern is also a clue; any pond spreadwing seen in late May or early June in our region is likely this species, but don't base identity on season alone. For additional tips on identifying females, see *Identification Clues* for Northern, Slender, and Lyre-tipped Spreadwings.

Life Cycle & Behavior: Flight period is short and emergence is early in our region, beginning in May and extending only into June. Adults are rarely seen here after mid-June. It is not known if this species breeds successfully in our region, or if adults seen here move up from the south. Eggs are laid into standing reed or rush stems above water level. Early life history of this species is not well-studied anywhere near our area.

Sweetflag Spreadwing *Lestes forcipatus*

Breeding: Well-vegetated or boggy ponds, pools and marshes, often temporary and lacking fish; occasionally in slow streams.

Nature Notes:

Species name means "bearing forceps," perhaps alluding to the shape of the male's cerci.

Emerges earlier in season than the similar Northern Spreadwing.

Distribution in our region is somewhat uncertain because of long confusion of this species with Northern and Southern spreadwings.

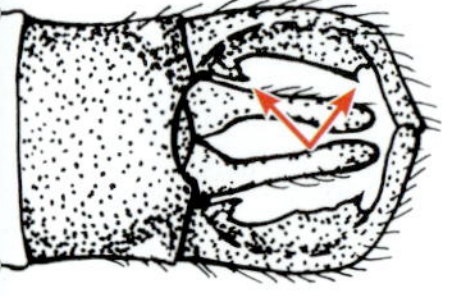

male's claspers; two unequal teeth on cerci

Description: Adults average 1.6 inches long. A mid-sized, dark spreadwing. Thorax and abdomen of male dark above with pale yellow or grayish-white sides. Lower side of thorax may have a black spot (↑) but this is not always visible. Thorax has pale or dark shoulder stripes. Dark coloration on thorax becomes more extensive with age and is eventually obscured by pruinosity. Last two abdominal segments are grayish-white pruinose. Wings clear. Female similar, but sides of her thorax are often light tan and her abdomen thicker. Rear of female's head can be dark or pale. Female's ovipositor is long, extending beyond the tip of her abdomen (↑).

Identification Clues: Separate male from male Northern Spreadwing by Sweetflag's two unequal teeth on cerci, end tooth is smaller and blunter than basal tooth; also, teeth are further apart (pg. 43). Further, Sweetflag not

The female Sweetflag Spreadwing is easy to spot because of her long ovipositor.

Note the bright blue eyes that are typical of most male spreadwings.

entirely pruinose on top of segment 2, showing shiny area at apical third; mature male Northern usually all pruinose there. Also, apical notch on segment 10 wide in Sweetflag, narrower, looking more pinched in Northern. See cautions on Southern Spreadwing page (pg. 50) about the very similar male. Female is unmistakable because of her very long ovipositor; only the female Emerald Spreadwing has an ovipositor nearly as long, but she is bright metallic green (not dark) on top of her thorax and abdomen.

Life Cycle & Behavior: Adults emerge in June and July and fly into September. Eggs are laid into stems of Sweet Flag (genus *Acorus*), bulrushes, and spikerushes above water level. Eggs hatch in early spring, at which time the pronymph emerges from the cut in the stem and quickly molts into the second instar. Nymphal development through about ten instars takes less than two months. Overwinters as egg that can withstand air temperatures to about minus 4 degrees F, but snow cover is needed to protect eggs from very cold temperatures.

Lyre-tipped Spreadwing *Lestes unguiculatus*

Breeding: Small, sunny ponds and marshes in open areas, often temporary in some years. Avoids acidic habitats and those with fish.

Nature Notes:

Six or more pairs in tandem have been seen laying eggs in a single bulrush stem.

Species name means "small-clawed," referring to the distinctly curved paraprocts (see illustration below).

Not an aggressive species; males may be chased by smaller damselfly species.

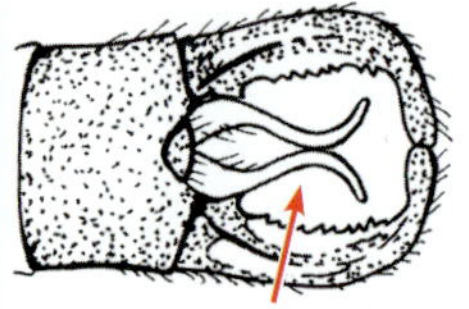

male's claspers; note lyre-shape of paraprocts

Description: Adults average 1.5 inches long. Thorax of male is bronze with pale green shoulder stripes (↑), turning pale yellow to whitish below. Thorax becomes increasingly pruinose with age. Abdomen metallic green, with last two segments grayish-white pruinose and segment 8 partially pruinose, often showing a dark basal triangle (↑). Wings clear. Female's thorax is similar, but her abdomen is dark with greenish reflections on top and thicker. Rear of female's head often with extensive pale area that reaches eyes. Female's ovipositor is short, not extending beyond the tip of her abdomen.

Identification Clues: Male's S-shaped paraprocts (↑) with divergent tips are diagnostic, however, be aware that paraprocts of other male spreadwings could be crossed, giving S-shaped look. Dark triangle on top of male's abdominal segment 8 is a fairly good field mark, but there

Note the male's pale green shoulder stripes. The male's S-shaped paraprocts form the lyre in Lyre-tipped. Also note for the unique dark mark atop segment 8 (right).

is some variation. Female Northern, Southern, and Slender spreadwings are similar to Lyre-tipped, but they have a darker rear of head not pale to the eyes, and all of them have segment 7 longer relative to the length of the ovipositor than Lyre-tipped. Positive identification is based on subtle differences in lengths of some abdominal segments measured under a microscope. Since male is easy to identify, look for males with which females are associated.

Life Cycle & Behavior: Adults emerge from mid June through early July and fly into September. Eggs are laid into emergent plant stems, often bulrushes or sedges, from two inches to two feet above water level. Eggs hatch in spring when water temperatures reach about 50 degrees F. Nymphal development takes about 60 days and maturation requires 16 to 18 days after emergence. Overwinters as eggs that can withstand air temperatures to about minus 4 degrees F. Snow cover is essential to protect eggs from very cold temperatures.

Slender Spreadwing *Lestes rectangularis*

MAY	JUNE	JULY	AUG	SEPT	OCT

Breeding: Shallow and partially shaded still-water habitats: temporary and permanent ponds, marshes and backwaters of slow streams.

Nature Notes:

Male is readily identified at a glance by his peculiarly long and thin abdomen.

Female lays eggs alone, without male in tandem or guarding nearby.

Prefers shadier shoreline areas than other pond spreadwings.

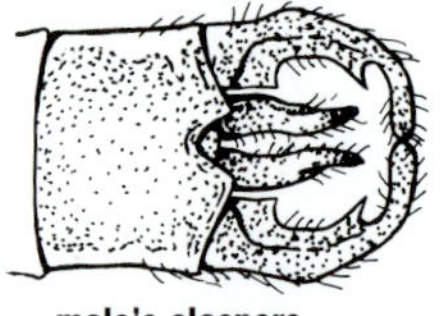

male's claspers

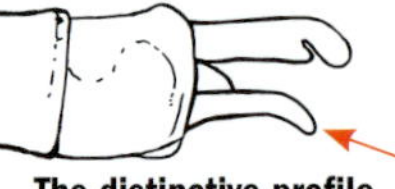

The distinctive profile of the male's claspers

Description: Adults average 1.8 inches long. A long, slender spreadwing. Thorax and abdomen of male dark above with pale yellow sides. Thorax has wide, bluish shoulder stripes that are often partly gray. Lacks pruinosity at tip of abdomen. Wings clear with pale vein at outer tip (↑), most easily seen on male. Female is similar, but her abdomen is shorter and stouter. Female ovipositor short, not extending beyond tip of abdomen, with bottom half dark.

Identification Clues: Male is easily identified by an abdomen that is twice as long as his wings (↑) with no pruinosity at the tip, and the downturned tips of his paraprocts seen in profile (↑) (see illus. this page). Female is similar to several other spreadwing females including Northern, Southern, and Lyre-tipped (see Identification Clues of those species). Her tarsi are usually all or partially pale, whereas females

Female also has a long slender abdomen and her wide, pale shoulder stripe often shows various amounts of blue & gray. Her ovipositor does not extend beyond tip of abdomen.

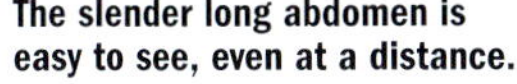
The slender long abdomen is easy to see, even at a distance.

Note the gray area on the pale shoulder stripe of this immature.

of the other similar species have entirely dark tarsi, and her abdominal segment 7 is relatively longer than on those species. Further, she oviposits alone, unlike our other pond spreadwings and her pale vein at wing tip might be visible against dark background. Positive identification of females of these species is based on subtle differences in lengths of some abdominal segments relative to the length of the ovipositor, measured under a microscope.

Life Cycle & Behavior: Adults emerge from mid June through July and often fly well into September. Eggs are laid into a variety of plant stems, including bulrushes, sedges and cattails, at water level or above. Eggs hatch in spring, at which time the pronymph emerges from the cut in the stem and quickly molts into the second instar. Nymphal development takes about twelve weeks and maturation requires about three weeks after emergence. Overwinters as a diapausing egg.

Emerald Spreadwing *Lestes dryas*

♂

U

MAY	JUNE	JULY	AUG	SEPT	OCT

Breeding: Small forest pools that dry up by summer (vernal pools); occasionally permanent ponds. Not where aquatic predators are common.

Nature Notes:

Species name refers to a mythical Greek wood nymph, perhaps indicating its preference for shaded, forest pools.

Thrives in shallow, temporary, woodland pools that lack fish.

Ranges across Europe and Asia as well as North America.

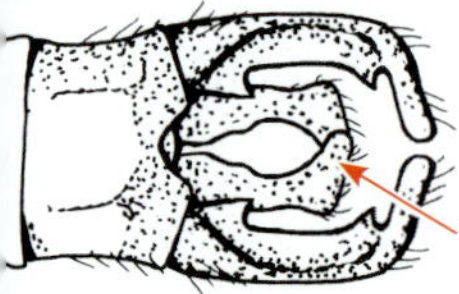

male's claspers; boot-shaped paraprocts

Description: Adults average 1.5 inches long. One of our most striking looking spreadwings and the stockiest. Thorax and abdomen of male is bright metallic green above (↑) and pale yellow turning whitish pruinose below (↑). Pale shoulder stripe very thin or absent. Last two abdominal segments are grayish-white pruinose. Wings clear. Female is also metallic green (↑), but with very thin, pale shoulder stripes and a thicker abdomen. She lacks pruinosity. Female has a long ovipositor that extends to or beyond tip of abdomen (↑) (almost as long as Sweetflag Spreadwing female).

Identification Clues: Boot-shaped paraprocts (widened at tips) of male are diagnostic (see

Female shows the same bright green top of thorax and abdomen as the male, but note her very long ovipositor that extends to or beyond the tip of the abdomen. Older males become increasingly whitish pruinose on the lower thorax (inset opposite page).

illustration opposite page). Brighter green and heavier bodied than other spreadwings. Female Sweetflag Spreadwing has an even longer ovipositor and the top of her thorax is dark, not bright green. Note, however, that tenerals of many pond spreadwings show metallic green in areas that are dark when mature.

Life Cycle & Behavior: Adults emerge in June and early July. Flight period usually ends by early August. Eggs are laid into emergent plant stems, often bulrushes, sedges, and grasses, above water. Eggs are large, hence the need for the female's large ovipositor. Eggs hatch in early spring after pools fill with spring run-off. The nymphs develop extraordinarily quickly through ten instars in just a few months, so they can emerge before vernal pools dry up. Overwinters as egg in late stages of embryonic development where they can withstand air temperatures lower than minus 4 degrees F. May overwinter as nymph in milder climates.

Swamp Spreadwing *Lestes vigilax*

Breeding: Wide variety of permanent waters including marshy and bog-bordered ponds, small acidic lakes and slow streams.

Nature Notes:

Species name means "watchful." Only Hermann Hagen, who named this species in 1862, knew why it is more watchful than others.

Often abundant where it occurs.

Mainly found during the second half of summer.

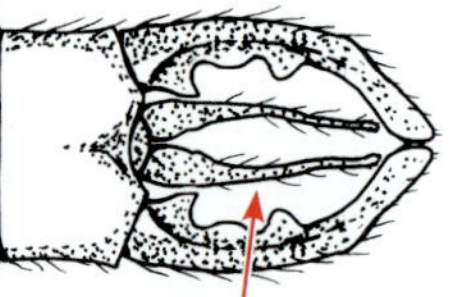

male's claspers; note long thin paraprocts

Description: Adults average 2 inches long. Large and slender bodied. Thorax and abdomen of male is metallic green above and pale yellow turning with age to whitish pruinose below (↑). Thin reddish-brown shoulder stripes (↑) usually visible. Last two abdominal segments are grayish-white pruinose. Wings clear. Female is similar to male overall, but her abdomen slightly shorter and thicker. Female may not always be distinguishable from female Elegant Spreadwing in the field.

Identification Clues: Terminal appendages of male are unmistakable with paraprocts long, thin and straight (see illustration this page). Distinguish females from Amber-winged by Swamp's clear wings, lack of oblique dark bands on side of thorax and angular basal plate of ovipositor. Distinguish females from Elegant Spreadwing by dark lower legs (↑), dark back of

Like many spreadwings, the female Swamp Spreadwing closely resembles the male. But note female's dark lower legs, dark back of the head and reddish-brown shoulder stripes, all which help to separate her from female Elegant Spreadwing.

head (yellow there on Elegant), and reddish-brown shoulder stripes. Also, lower half of ovipositor usually pale (dark on Elegant), and basal plate with pointed tooth (angular on Elegant). Further, top of abdominal segment 10 metallic green on Swamp, pale or pruinose (not metallic) there on Elegant.

Life Cycle & Behavior: Adults emerge in June and July and may fly into September. Surprisingly, little life history information is available about this widespread and fairly common species. Eggs are laid into emergent plant stems above water. The nymphal period spans at least nine months. Overwintering occurs during one of the later instars.

Elegant Spreadwing *Lestes inaequalis*

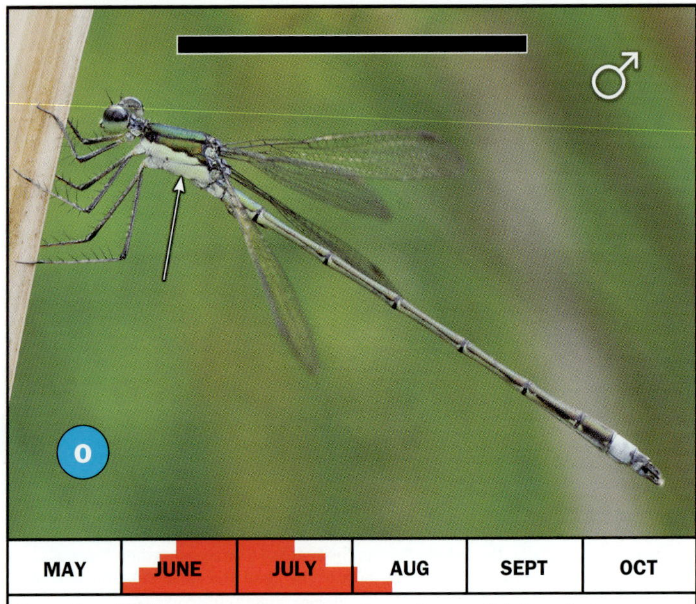

Breeding: Wide variety of permanent ponds, small lakes, lagoons and slow streams.

Nature Notes:

Elegant is a great name for this large, handsome and slender spreadwing.

Species name means "unequal", referring to the paraprocts of the male being distinctly longer than the cerci (illustration below).

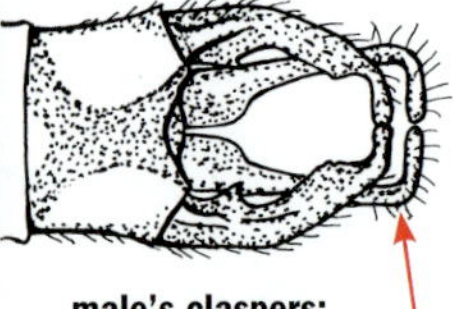

male's claspers; paraprocts are longer than cerci

Description: Adults average 2.1 inches long. Large, attractive, slender bodied pond spreadwing. Thorax and abdomen of male is bright green above and pale yellow below (↑). Very thin, dark shoulder stripe may be visible. Last two abdominal segments are grayish-white pruinose. Pruinosity on thorax develops less fully than on other spreadwings. Wings clear. Female is similar to male, but her dark shoulder stripe usually wider than on male. Her abdomen with all segments metallic green above, including segment 10. Female may not always be distinguishable in the field from Swamp Spreadwing, but in the hand many differences can be seen.

Identification Clues: Terminal appendages of male are unmistakable with paraprocts longer than cerci (see illustration this page). Distinguish females from Amber-winged

Female has pale yellow on lower tibiae. The female Elegant Spreadwing is very similar to the male but not quite as bright. Both show pale yellow on lower part of thorax.

Spreadwing by Elegant's clear wings, lack of oblique dark bands on side of thorax and angular basal plate of ovipositor. Distinguish females from Swamp Spreadwing by Elegant's pale yellow on lower legs (tibiae) (↑) and back of head (see also Swamp Spreadwing *Identification Clues*).

Life Cycle & Behavior: Adults emerge in June and July and may fly into August. Life history of this species has not been thoroughly studied. Eggs are laid into emergent plant stems above water. Overwintering likely occurs during one of the last three nymph instars.

Amber-winged Spreadwing *Lestes eurinus*

Breeding: Small lakes and ponds, usually permanent and usually lacking fish.

Nature Notes:

One of our largest pond spreadwings and one of the earliest to appear.

Species name refers to the east wind; no one seems to know how this applies to the insect.

Film of pruinosity often give the thorax a bluish cast.

Flies actively along shorelines and over open water.

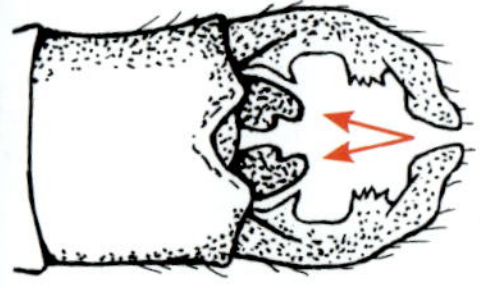

male's claspers; paraprocts much shorter than cerci

Description: Adults average 2 inches long. Large and stout-bodied. Head, thorax and abdomen of male metallic green above, thorax abruptly turning to pale yellow below with a pair of dark oblique bands, one at base of hind leg, the other behind it. The last two abdominal segments are grayish-white pruinose. Film of pruinosity on thorax develops with age obscuring much of the underlying color pattern and giving a bluish hue. Wings entirely covered with an amber wash (↑). Female is similar, also showing two dark bands on lower thorax (↑).

Identification Clues: Honey-colored wings, dark bands on thorax, and large size readily separate both sexes of this species from all other spreadwings. Shapes and relative sizes of terminal appendages of male are diagnostic, with cerci more than twice as long as paraprocts (see illustration this page). With females in

♀

The heavy-bodied female is one of the most robust damselflies in our region (above). With age, a film of pruinosity obscures the male's thoracic marks (below).

female abdomen tip

♂

the hand, also look for rounded margin of basal plate of ovipositor (illustration above).

Life Cycle & Behavior: Adults emerge in June and July and may fly through August. After emerging, females may require up to a month to mature. American Bur-Reed (*Sparganium americanum*), along with a variety of rushes and sedges, are favored plants for egg laying. Eggs are laid into stems above water and take about 45 days to develop. Upon hatching, the pronymph emerges from the cut in the stem, falls into the water and molts into the second instar within three minutes. The nymphal period spans about nine months and includes 13 to 17 instars. Overwintering occurs during one of the last three nymph instars.

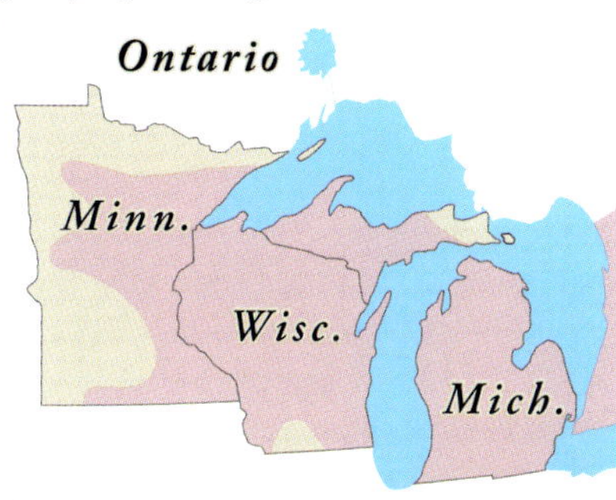

Pond Damsels
Family Coenagrionidae

The pond damsels comprise a huge family of mostly small, clear-winged species that delightfully display nearly all the colors of the rainbow. This family includes our smallest damselflies, some of which barely reach an inch long, and many of our most common species. Aspects of wing venation are the unifying features that set them apart from other families (see Family Key on pg. 29). Their stigmas are short, less than twice as long as wide. The male's terminal appendages are shorter than in other families, and the legs are quite short as well. Females often come in two color schemes: a less-common male-like color pattern (homeochromatic) and a typical color pattern that is different than the male (heterochromatic).

Pond damsels inhabit the full spectrum of aquatic breeding habitats in our region: lakes, ponds, marshes, swamps, bogs, fens, rivers, streams and spring seeps. Some species are habitat generalists that can be found at almost any aquatic site whereas others are closely tied to specific types of habitat (specialists). Adults can often be found near breeding sites around clearings, meadows, along roadways and railroad grades. They fly from May through September in our region, although most species have a short flight period lasting only four to six weeks.

Worldwide there are over 1,140 described species of pond damsels, with about 105 species in North America north of Mexico. This is easily the largest damselfly family in our region, with 36 species addressed in this guide.

Eurasian Bluets — Genus *Coenagrion*

The Eurasian bluets are a northern group of small to medium-sized damselflies that I find to be exceptionally beautiful. Males of the Taiga Bluet are especially compelling, with shades of blue and green on their thorax and abdomen that are hard to adequately describe. Males of most species are primarily blue and black and can easily be mistaken for male American bluets (genus *Enallagma*) or even forktails (genus *Ischnura*). Males are distinguished from these genera by differences in the shapes of the terminal appendages, and also the colors and patterns of the thorax and abdomen. Females show either of two basic color patterns: male-like hues (homeochromatic) and duller, greenish or brownish hues (heterochromatic). Females can be separated in the hand from American bluets by their lack of a vulvar spine on abdominal segment 8 (all American bluet females have it).

Adult Eurasian bluets tend to fly low and perch often among vegetation. Our northern species are usually found along the edges of ponds, marshes and bogs. Males arrive at breeding sites before females. They are rarely aggressive with other damselflies and do not defend territories. Females oviposit in tandem on emergent or floating vegetation with males alertly guarding in the "sentinel" position.

Nymphs live in still, often shallow waters where they crawl about on submerged vegetation hunting their prey. The nymphs are small and brown or greenish and are very similar to nymphs of American bluets. North American species generally complete their life cycle in one year, but in the Far North development is slower, possibly taking as long as three or four years. This genus is better able to tolerate colder climates than any other damselfly group. A fascinating study found that the nymphs of the Taiga Bluet and the Prairie Bluet overwintered embedded in five to eight inches of ice! They were often found completely encased in an upside-down position as though trapped while walking about on the undersurface of the ice, and when the ice melted they resumed activity. The ability to overwinter in this way allows them to survive in shallow northern ponds that freeze to the bottom in winter and protects them from predators.

This is a circumpolar boreal group that occurs around the world, with most species occurring in Europe and Asia. Only three species are found in North America, all of which are found in our region. The Taiga Bluet is fairly widespread and common, though rarely abundant at any site. The Subarctic Bluet is one of our rarest damselflies. Keep your eyes open also for the Prairie Bluet, which occurs along the western edge of our region and might be more widespread throughout Minnesota than currently thought.

Eurasian Bluets — Genus *Coenagrion* (Species begin on page 70)

American Bluets — Genus *Enallagma* (Text on page 76)

Forktails — Genus *Ischnura* (Text on page 111)

Red Damsels — Genus *Amphiagrion* (Text on page 125)

Aurora Damsel — Genus *Chromagrion* (Text on page 127)

Sprites — Genus *Nehalennia* (Text on page 128)

Abdomen tip side views of male Pond Damsels

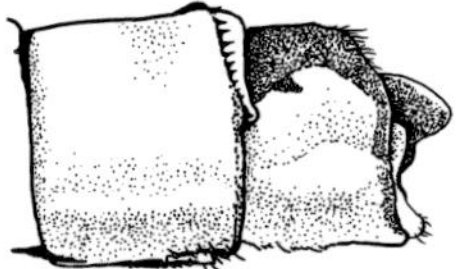

Prairie Bluet
Coenagrion angulatum pg 70

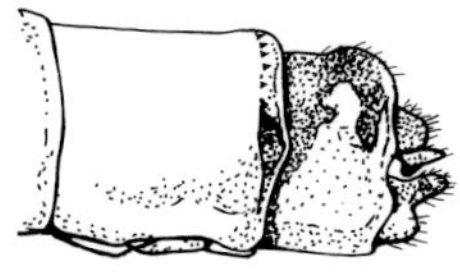

Subarctic Bluet
Coenagrion interrogatum pg 72

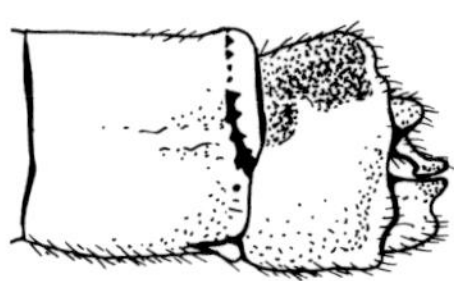

Taiga Bluet
Coenagrion resolutum pg 74

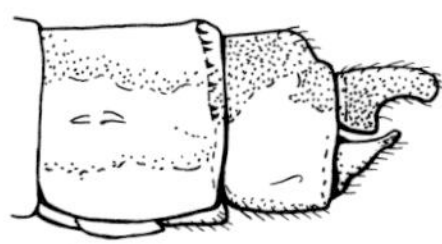

River Bluet
Enallagma anna pg 78

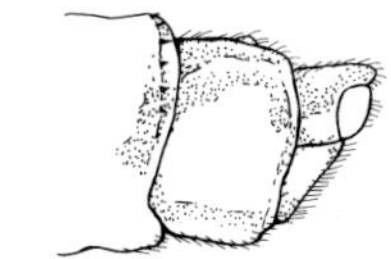

Familiar Bluet
Enallagma civile pg 80

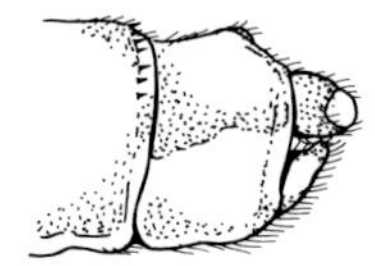

Tule Bluet
Enallagma carunculatum pg 82

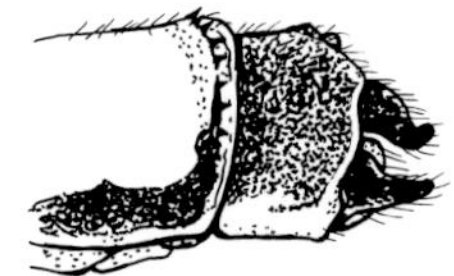

Skimming Bluet
Enallagma geminatum pg 84

Azure Bluet
Enallagma aspersum pg 86

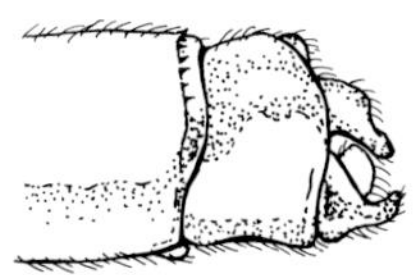

Alkali Bluet
Enallagma clausum pg 88

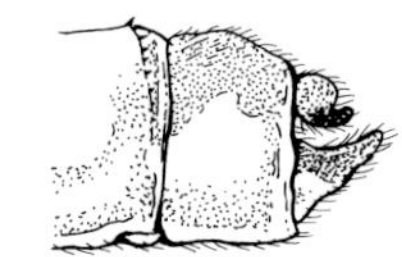

Northern Bluet
Enallagma annexum pg 90

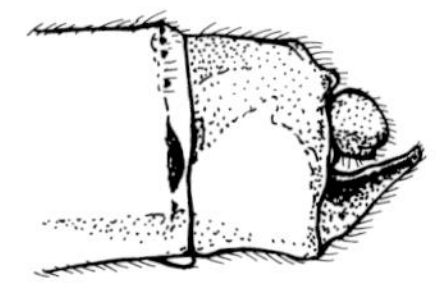

Boreal Bluet
Enallagma boreale pg 92

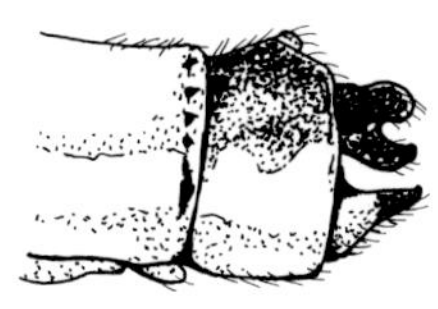

Marsh Bluet
Enallagma ebrium pg 94

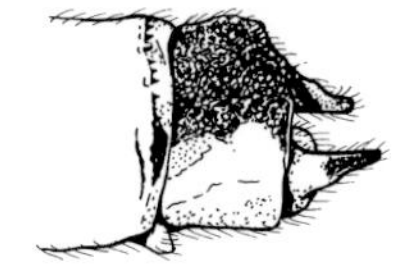

Hagen's Bluet
Enallagma hageni pg 96

Stream Bluet
Enallagma exsulans pg 98

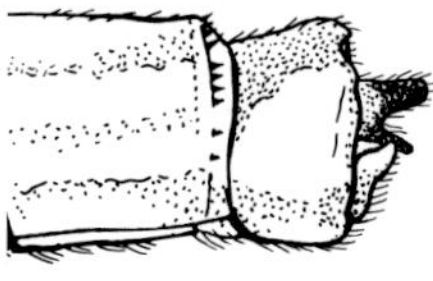

Rainbow Bluet
Enallagma antennatum pg 100

Slender Bluet
Enallagma traviatum pg 102

Double-striped Bluet
Enallagma basidens pg 104

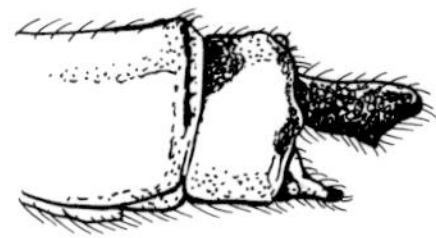

Orange Bluet
Enallagma signatum pg 106

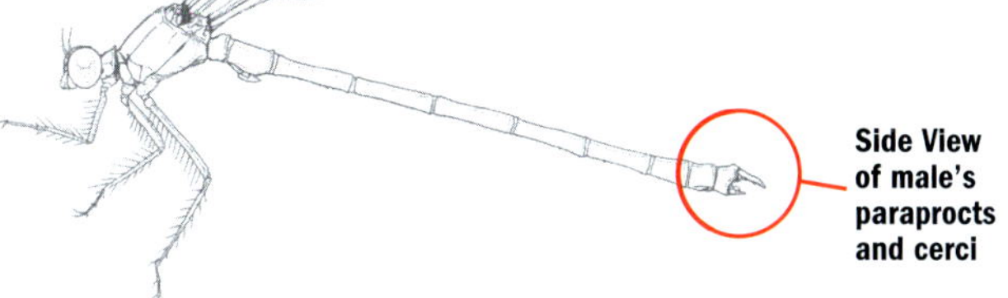

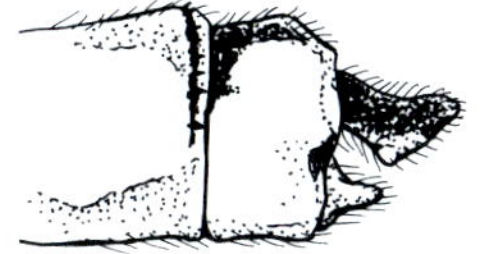

Vesper Bluet
Enallagma vesperum pg 108

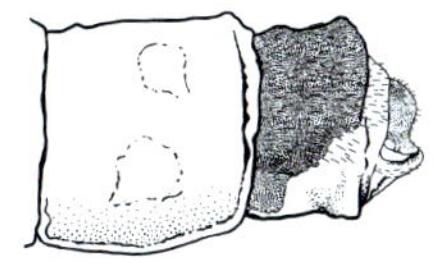

Turquoise Bluet
Enallagma divagans pg 110

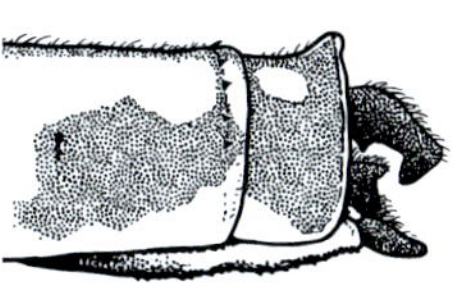

Lilypad Forktail
Ischnura kellicotti pg 112

Plains Forktail
Ischnura damula pg 114

Western Forktail
Ischnura perparva pg 116

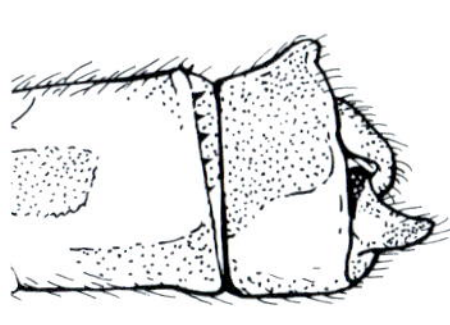

Eastern Forktail
Ischnura verticalis pg 118

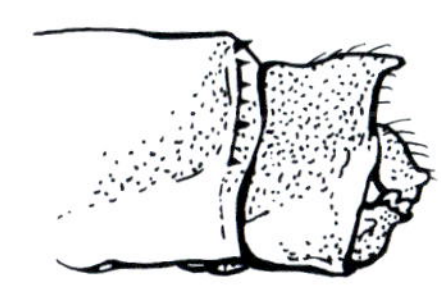

Fragile Forktail
Ischnura posita pg 120

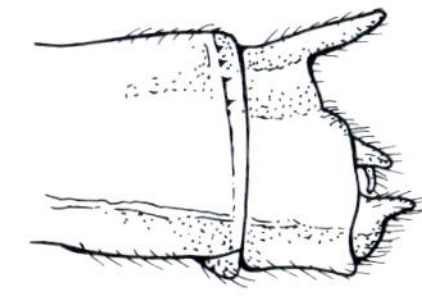

Citrine Forktail
Ischnura hastata pg 122

Western Red Damsel
Amphiagrion abbreviatum pg 124

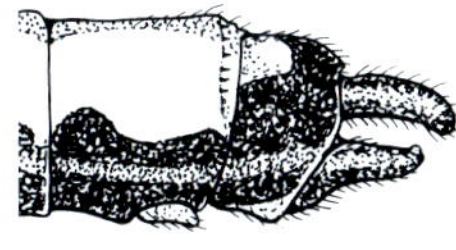

Aurora Damsel
Chromagrion conditum pg 126

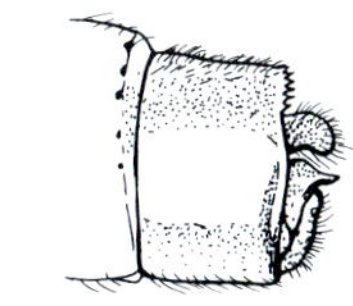

Sphagnum Sprite
Nehalennia gracilis pg 129

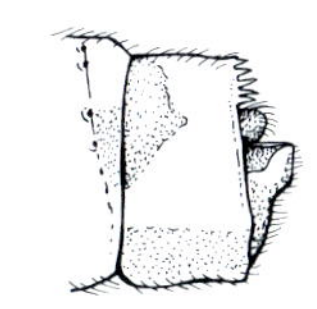

Sedge Sprite
Nehalennia irene pg 130

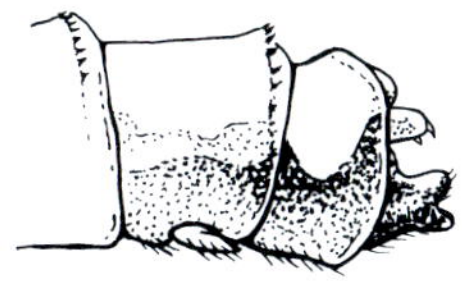

Blue-fronted Dancer
Argia apicalis pg 134

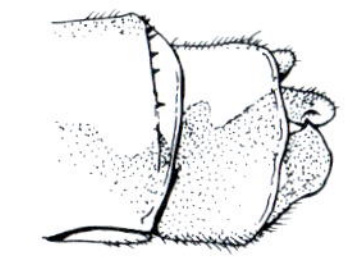

Powdered Dancer
Argia moesta pg 136

Blue-ringed Dancer
Argia sedula pg 138

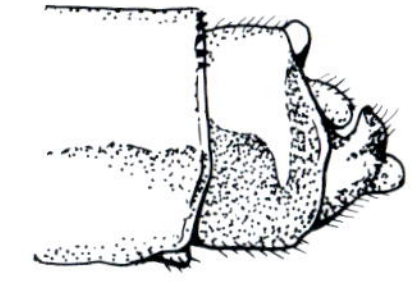

Variable Dancer
Argia fumipennis pg 140

Springwater Dancer
Argia plana pg 142

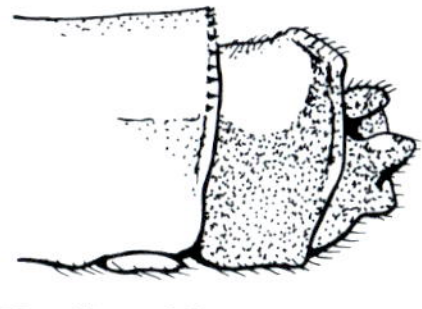

Blue-tipped Dancer
Argia tibialis pg 144

Prairie Bluet *Coenagrion angulatum*

Breeding: Wide variety of still-water habitats in open (prairie) landscapes; sometimes slow streams.

Nature Notes:

Can be very abundant around favored prairie wetlands.

Males avoid flying over open water, instead remaining in grasses near shore.

A northern and western species that is common in sunlit ponds and sloughs of the Canadian prairie. Look for it along the western edge of our region.

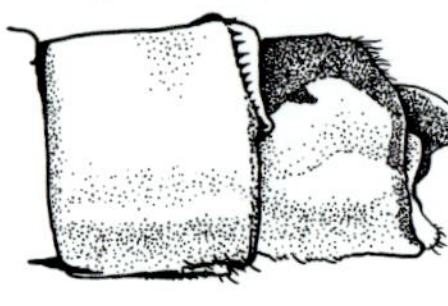

male's abdomen tip

Description: Adults average 1.2 inches long. Male is a small black-type bluet colored much like several American bluets. The thorax is blue and black with dark shoulder stripe about same width as blue stripe. Male's mid-abdominal segments black with thin, blue rings at bases (↑); blue on segments 1 and 2, and 8 and 9. Female's thorax like male's, but pale areas may be light greenish or blue. Female's abdomen nearly all black above (↑) with very thin, pale rings at bases that are a bit wider on segments 8 and 9 (↑); on segment 8 the ring is often divided into two basal marks. Eyespots of both sexes mid-sized.

Identification Clues: Male is slightly larger than male Skimming Bluet and lacks the wavy black mark on segment 2 of that species. Male could be mistaken for several other black-type American bluets in the field, but habitats

The top of this female's abdomen is almost totally black. Note that the segment 8 ring is often divided into two basal marks.

usually different and ranges may not overlap; in the hand Prairie's terminal appendages are diagnostic (see illustration on opposite page), not like those of any American bluet. Best "tells" for female are nearly all-black abdomen top with pair of pale marks at base of segment 8, and lack of vulvar spine.

Life Cycle & Behavior: Adults emerge in June and sexual maturation takes about one week. Mating occurs near the site of maturation without courtship or territorial behavior. Oviposition is usually in tandem and occurs into plant stems below the water's surface. Overwinters in nymph diapause while embedded in ice in one of the last three instars.

Subarctic Bluet *Coenagrion interrogatum*

Breeding: Cold swamps and open bogs, especially quaking sphagnum moss margins around bog ponds.

Nature Notes:

One of the rarest damselflies in our region.

The southern limit of its distribution is further north than any other damselfly.

Found at Churchill on Hudson Bay and northwest to the Yukon Territories.

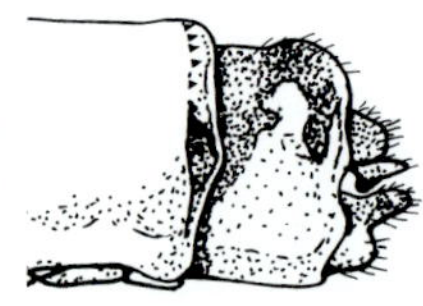

male's abdomen tip; shape of cerci and paraprocts are unique

Description: Adults average 1.2 inches long. Both sexes are a study in blue and black. The pale shoulder stripes on the thorax are wide and bordered and divided with black, which forms a pale blue square and rectangle (↑) on the shoulder. The abdomen is blue and black, with the patterns on segment 2 unlike any other species. Pale eyespots are prominent on the otherwise dark head. Wings clear.

Identification Clues: Males have unique dark markings including the pale shoulder stripe feature described above and the underside of the thorax, which is extensively black including a Y-shaped figure. He has more blue at the tip of the abdomen than any other bluet in our region except male Azure Bluet. The shape of the male's terminal appendages is diagnostic (see illustration this page). The female's thorax undersides are heavily marked with black and

The female Subarctic Bluet is extensively marked with black on her lower thorax and sides of her abdomen.

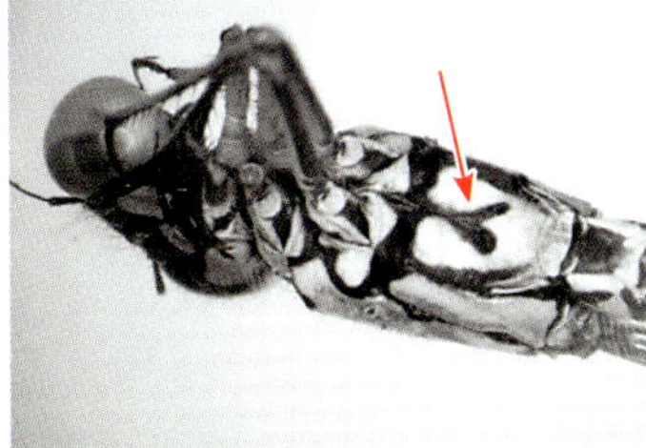

Note distinctive Y-shaped mark on underside of thorax of both sexes (left); this is only visible on specimens in the hand.

Look for pale blue square and rectangle on shoulder (right).

she shows long black streaks on the sides of her abdomen (↑).

Life Cycle & Behavior: Flight season is early and relatively short, with emergence beginning in mid May to early June. Females lay eggs in tandem or alone into various leaves and stems near water level throughout the month of June. Sexual maturation takes about one week. Oviposition has not been described and the early nymph stages are poorly known. Likely overwinters in one of the last three nymph instars.

Taiga Bluet *Coenagrion resolutum*

Breeding: Wide variety of still waters from marshy ponds and grassy ditches to bogs, fens and slow streams.

Nature Notes:

Ranges above the Arctic Circle in Canada—all the way to the Arctic Ocean!

Nymphs overwinter embedded in ice but do not freeze solid!

Species name means "resolute," underscoring the fortitude they possess to eke out an existence in inhospitable northern climates.

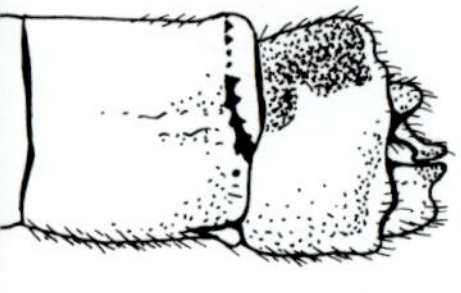

male's abdomen tip

Description: Adults average 1.2 inches long. Male is an attractive combination of blue, green and black. His thorax is blue with a wide black top stripe and thinner black shoulder stripes. Sides of thorax blue, shading to green below (↑). Abdomen is blue and black; segment 2 with a U-shaped dorsal black mark (↑). Segments 6 and 7 mostly black, 8 and 9 entirely blue. Female's head and thorax similar, but with top of her abdomen almost entirely black. Pale areas are either blue like male or yellowish green. Wings clear.

Identification Clues: Identify males by blue-green color on side of thorax, and unique pattern of black on abdomen including all black segments 6 and 7 and U-shaped black mark on top of segment 2. Net some males to confirm identity by diagnostic shape of terminal appendages in side view (see illustration this

♀

The female Taiga Bluet resembles several female American bluets but lacks a vulvar spine on abdomen segment 9. Top of female's abdomen almost entirely black.

page). In females, look for blue-green sides on thorax (↑) and the lack of a vulvar spine. Use also as a clue the tendency of both sexes to perch with wings slightly spread.

Life Cycle & Behavior: Adults emerge from late May to late June when water temperature exceed 54 degrees F. Sexual maturation takes about one week. Pairing and egg laying start in early June and adults may fly late into July. Mating occurs at the site of maturation, about 100 yards away from the pond, without courtship or territorial behavior. Oviposition is always in tandem and occurs below the surface into rushes, cattails, pondweed and other standing or floating aquatic plants. Eggs hatch within three weeks. Growth of nymphs extends for ten to 22 months. Overwinters in nymph diapause and embedded in ice in one of the last three instars.

American Bluets
Genus *Enallagma*

The American bluets are a large genus of small to mid-sized species that are abundant in our region around a wide variety of still-water habitats. They are commonly called "bluets" because the males of most species are primarily blue with contrasting black markings that consist of distinct stripes on the thorax and bands on the abdomen. While glancing through the field notebook of an Ontario friend a few years ago I saw reference to "lots of BBBs" and asked him what that stood for. "Black and Blue Boys," he replied with a smile, referring to the males of this genus. Unlike many damselflies, males remain colorful throughout life, as there is little tendency for the brilliant blues to darken or become obscured with pruinosity as they age. Males can sometimes be identified in the field by their color patterns, but firm identifications often require examination in the hand, especially for blue-type bluets (see below), where the distinctive shapes of their terminal appendages can be seen in side view with a hand lens (pgs. 68-69). Even when color patterns seem sufficient to identify a species, it is always a good idea to confirm it by comparing the side view of the cerci and paraprocts with the drawings in this guide. If you get into the habit of doing this early, you won't regret it!

To ease the task of identification with this large genus, it is helpful to categorize the males into subgroups. At the first division I separate the "citrus-hued" bluets (Orange Bluet and Vesper Bluet) that have the pale color yellow or orange, from the dominant groups of males having the pale color blue. Next, I use a system pioneered by Ed Lam of putting males with pale colors blue into three subgroups: "black-type" bluets have the tops of the abdomens mostly black (especially the middle segments 3 through 5); "intermediate-type" bluets have roughly equal amounts of black and blue on the middle abdominal segments; and "blue-type" bluets have the middle abdominal segments more blue than black. The Rainbow Bluet presents an exceptional case because of its green thorax and orange face (pg.

Black-type bluet **Intermediate-type bluet** **Blue-type bluet**

100). Female American bluets are duller than males in color, typically showing green, yellowish green or tan hues where the males are blue. Females of the blue-type bluets are often blacker on top of the abdomen than the males. Females of many species have male-like forms, which are usually less common than the typical heterochromatic form. Females can sometimes be identified in the hand, or even in the field, by distinctive color patterns, but some "clusters" of species are very similar, requiring careful examination of their mesostigmal plates under a microscope. Females of these species can often be identified (tentatively) by identifying the males they are associated with. Females have a vulvar spine, contrasting with the Eurasian bluets, which lack the spine. Wings of both sexes are clear, moderately "stalked," and have dark stigmas, unless noted differently in the species accounts. Eyespots are always present and are helpful in identification because they vary in size among species. The tibial spurs are short, in contrast with the dancers.

These familiar, bright-blue damselflies often abound in grasses and sedges around ponds and lakes in early summer. Well-vegetated still-waters are the habitats of choice for most species, but other habitats are sometimes used, including streams and rivers. Most American bluets are content to make short forays close to vegetation after tiny insect prey, although some do fly strongly over water. Males do not usually defend territories or show much aggression toward other males. Females oviposit usually while in tandem but sometimes alone, into living or rotting plant stems or algal mats, sometimes while fully submerged.

The nymphs are green or brown and of moderate build, with gills that are variable in shape, though usually narrow. Ponds without sunfishes or black bass (family Centrarchidae) favor species of bluets that escape by swimming away (Azure Bluet, Boreal Bluet and Northern Bluet). Our other bluets do just fine in waters with fish. These species do not try to escape from fish predators by out-swimming them (a fruitless undertaking), but hide instead. All species in our region overwinter as nymphs.

This worldwide genus has more species in North America (at least 37 of them) than are found anywhere else. A diverse group of 17 species inhabits our region. The phylogenetic status of a possible eighteenth species, the Vernal Bluet (*Enallagma annexum vernale*), is presently being studied and vigorously debated among odonatists. Because the Vernal Bluet intergrades extensively with the Northern Bluet in our region, and even experts have trouble telling them apart, I am considering them to be subspecies for now until more data about their morphology, habitat and genetic relationship become available.

River Bluet *Enallagma anna*

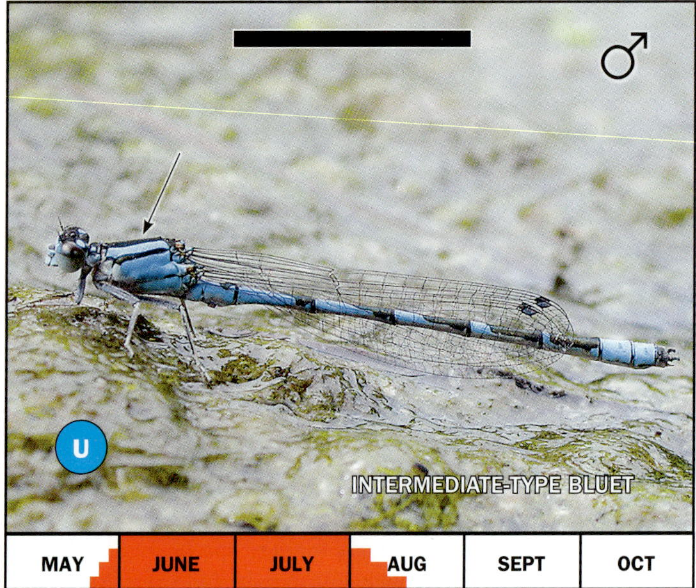

MAY	JUNE	JULY	AUG	SEPT	OCT

Breeding: Slow-flowing small to medium-sized streams with gentle to moderate flow; sometimes irrigation ditches and warm springs.

Nature Notes:

A western species that appears to have recently extended its range to the east.

May hybridize with other closely related bluets.

Carefully examine intermediate-type bluets along streams in partly open landscapes in our region for this uncommon species.

Frequently perches on the ground.

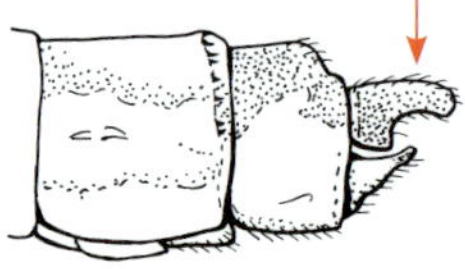

male's abdomen tip; note long cerci

Description: Adults average 1.3 inches long. An intermediate-type bluet of flowing waters. Male has a blue and black thorax with a dark shoulder stripe narrowing rearward (↑). Male's middle abdominal segments about half black, half blue, 6 and 7 mostly black and segments 8 and 9 all blue. His eyes black above, blue below. Female shows pale areas of blue, brown or olive with narrow dark shoulder stripes. Dark markings on top of her abdomen torpedo-shaped, pointing toward thorax. Both sexes have eyespots of moderate size (↑).

Identification Clues: Male is similar to other intermediate-type bluets with narrow dark shoulder stripes, but is readily identified in the hand by the long, distinctively shaped cerci (see illustration this page) that are sometimes visible in the field or in photographs. Most likely to be mistaken in the field for other intermediate-

Female River Bluets may be all blue, as this one, or partially or entirely olive.

Pair of River Bluets ovipositing in a stream. Female is depositing eggs while the male contact-guards her.

type bluets like Alkali, and especially, Tule. Female is very similar to Boreal and Northern Bluets and can only be identified with certainty under a microscope. As usual, note identity of males with which females are associated.

Life Cycle & Behavior: Little is known about the ecology and reproduction of this species in our region. Adults fly primarily in June and July. Females oviposit singly or in tandem into floating vegetation and plant stems. The nymphs are unusual among bluets because they live in streams (along with Stream Bluet) where they overwinter in later instars.

Familiar Bluet *Enallagma civile*

Breeding: Wide variety of still-water habitats including lakes, ponds, marshes and bogs. Even in brackish and alkaline waters.

Nature Notes:

This notable disperser is expanding its range north into newly created habitats like mitigation marshes.

Males are peaceful to a fault—they do not compete for territories, nor do they court females or engage in displays of any kind. Even the species name means "polite."

When female finishes ovipositing into an underwater stem, or is threatened by a predator, she lets go and pops to the surface.

male's abdomen tip; cerci have pale tubercles at tip

Description: Adults average 1.3 inches long. The male of this fairly large, blue-type bluet has a blue and black thorax with narrow, dark shoulder stripes. Male's abdomen is mostly blue on segments 2 through 5 (↑) with black areas increasing on 6 and 7; segments 8 and 9 are all blue. Eyes mostly blue. Female's pale areas are blue to olive-tan with her thorax markings similar to male's. The top of her abdomen is mostly dark, but with pale areas at the bases of some segments. Eyespots are moderately sized and tear-shaped (↑).

Identification Clues: Male is a bit larger than the Marsh and Hagen's Bluet and has less black on his mid-abdominal segments than the Tule, River, and Alkali bluets. Northern and Boreal Bluets have larger eyespots. Terminal appendages seen in side view are unmistakable with dorsal "arm" over pale

Female Familiar Bluets like this blue-form specimen are hard to separate from other female American bluets.

Note the tear-shaped eyespots and the blue shoulder stripe that is wider than the dark stripe.

tubercle (see illustration on opposite page). Female cannot be reliably distinguished from several other bluets except under a microscope.

Life Cycle & Behavior: The Familiar Bluet has a very long and apparently bi-modal, flight period, with emergence peaks in June or early July and again in late August and September. Some individuals may fly into October. Females oviposit into plant stems first in tandem above the surface and then alone beneath the surface. Nymphs live for nearly a year and overwinter in one of the later instars.

Tule Bluet *Enallagma carunculatum*

MAY	JUNE	JULY	AUG	SEPT	OCT

Breeding: Lakes, ponds and slow rivers, especially open shores of large lakes near bulrushes. Tolerates saline and brackish waters.

Nature Notes:

Often found at the open shores of larger, deeper lakes including the Great Lakes.

Tule Bluets have a long, mid-summer flight period in our region.

Known to hybridize with Familiar Bluet and River Bluet.

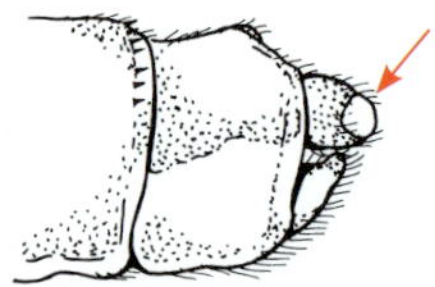

male's abdomen tip; cerci have pale tubercles that extend out from tip.

Description: Adults average 1.3 inches long. The male of this common, intermediate-type bluet has a blue and black thorax with a fairly narrow dark shoulder stripe of moderately even width (↑). Male's abdomen is about half blue on segments 2 through 4 (↑) with increasing amounts of black on 5, 6 and 7; segments 8 and 9 are all blue. Female's pale areas are blue to olive-tan with her thorax markings similar to male's. The top of her abdomen is mostly dark but with pale markings at the bases of some segments. Eyespots rather small, especially on the female (↑).

Identification Clues: Male shows more black on his mid-abdomen than blue-type bluets; because it is common in our region it can often be tentatively identified in the field by that character alone, but always confirm by examining terminal appendages in side

Female Tule Bluets may be olive, tan or blue. Note this blue-form female's tiny eyespots.

view which are distinctive (see illustration on opposite page). Each cercus has a pale tubercle at its tip. Has more black on segments 3 and 4 than River or Alkali Bluets and has smaller eyespots than Northern and Boreal Bluets. Female cannot be reliably distinguished from several other bluets except under a microscope.

Hazards are everywhere in our region. These two males are stuck to the flower head of a sow thistle.

Life Cycle & Behavior: This species emerges from mid June through July and flies into September. Emergence may occur high on rush stems, a foot above water. Females oviposit in tandem into stems of bulrushes and other plants that may be submerged, near surface, or well above water. Egg laying may occur into autumn. Nymphs live for nearly a year, have eight to 12 instars (usually ten) and overwinter in one of the later instars.

Skimming Bluet *Enallagma geminatum*

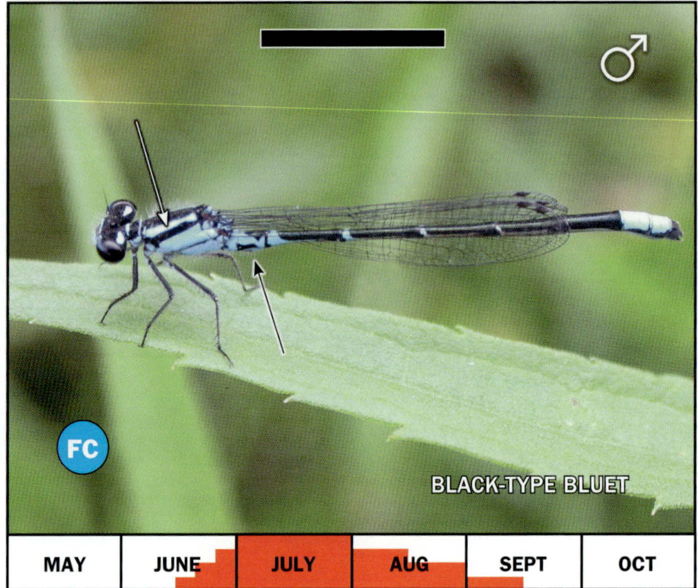

MAY	JUNE	JULY	AUG	SEPT	OCT

Breeding: Ponds, lakes and slow streams, usually clear and well-vegetated, especially with water lilies.

Nature Notes:

Flies low over lily pads and lands on them often, making them difficult to net.

Active in the middle of the day when light intensity is high.

Nymphs evidently need clean, well-oxygenated water, perhaps to a greater extent than most damselfly species.

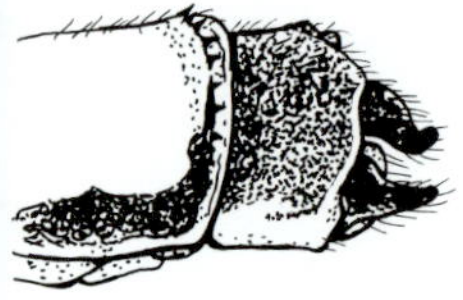

male's abdomen tip

Description: Adults average 1 inch long.

A tiny, unmistakable black-type bluet. Male has a blue and black thorax with a thick, dark shoulder stripe and a blue shoulder stripe that may be constricted near the middle (↑). Male's abdomen is mostly black above on segments 2 through 7 with segments 8 and 9 all blue. Female is similar but with a pair of blue spots that may be fused on top of segment 8 (↑). Both sexes have a wavy blue line bordered above and below with black on the side of abdominal segment 2 (↑).

Identification Clues: Their tiny size and the wavy blue line on the side of abdominal segment 2 set both sexes apart from all other damselflies in our region. Male is most like male Slender Bluet, but Slender lacks the segment 2 color pattern, has larger eyespots, and is larger. Male Lilypad Forktail has blue ring instead of wavy line on segment 2, larger eyespots, a wider

The female Skimming Bluet has a pair of blue spots on the top of abdominal segment 8.

Note the distinctive wavy blue stripe bordered by black on this male's abdominal segment 2.

dark shoulder stripe, and more blue at abdomen tip. Always wise to confirm identification by looking at shapes of terminal appendages in side view. Two blue dots on top of female's abdominal segment 8 are unique and often visible in the field.

Life Cycle & Behavior: An extended emergence period begins in June and adults fly throughout the summer, often into September. Females oviposit in tandem or alone, laying the eggs into emergent or floating vegetation including algae. Nymphs overwinter in one of the later instars.

Azure Bluet *Enallagma aspersum*

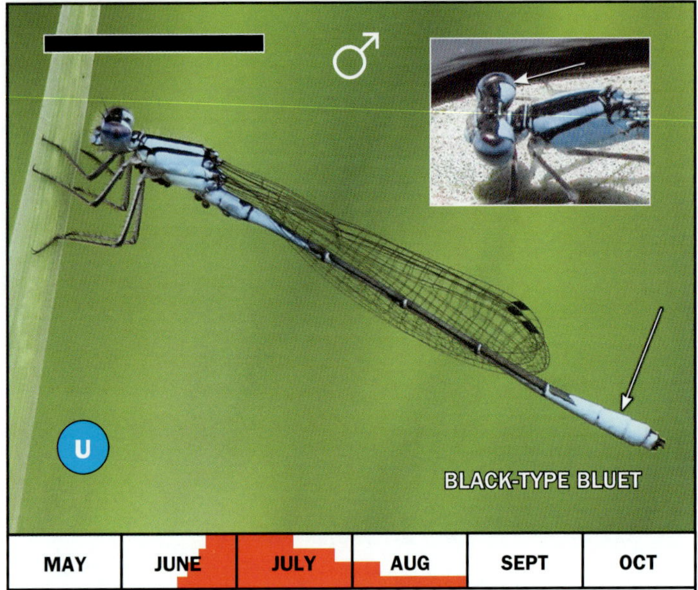

MAY	JUNE	JULY	AUG	SEPT	OCT

Breeding: Shallow ponds and small lakes, sometimes boggy and usually lacking fish, including temporary and newly formed ponds.

Nature Notes:

Local in our region, but may be abundant once found.

Males aggressively attempt to drive other males from choice breeding sites.

Like Familiar Bluet, quickly colonizes newly created habitats.

Nymphs sometimes found in ponds subject to periodic winterkill of fish.

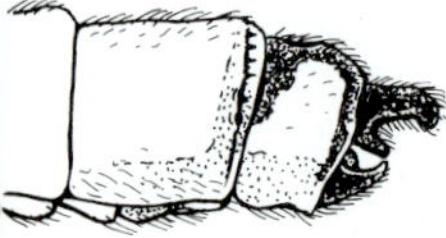

male's abdomen tip

Description: Adults average 1.2 inches long.

The male of this mid-sized, black-type bluet has a blue and black thorax with a narrow, dark shoulder stripe. Male's abdomen is mostly black above on segments 4 through 6, segments 8 and 9 are entirely blue and segment 7 is mostly blue (↑). Male's eyespots are very large (↑). The female is similar to the male, except for her abdomen, which has paired blue spots on top of segments 7 (large spots) and 8 (small spots) (↑).

Identification Clues: Both sexes are unmistakable. The male is only black-type bluet in our region with segment 7 usually at least half blue. Confirm male by examining terminal appendages in side view (see illustration this page). The female is the only bluet in our region having paired blue spots on top of segments 7 and 8.

The female Azure Bluet is one of the few American bluets to have only a blue form. Pairs of pale spots near the abdomen tip make these females easy to recognize in the field.

Life Cycle & Behavior: Adults have a fairly long flight period, emerging in June and flying through August. Females oviposit alone or in tandem into submerged vegetation, sometimes going alone quite deep. Females may remain submerged for 25 minutes. Eggs hatch in about three weeks. Nymphs live underwater for about a year and overwinter in one of the later instars.

Alkali Bluet *Enallagma clausum*

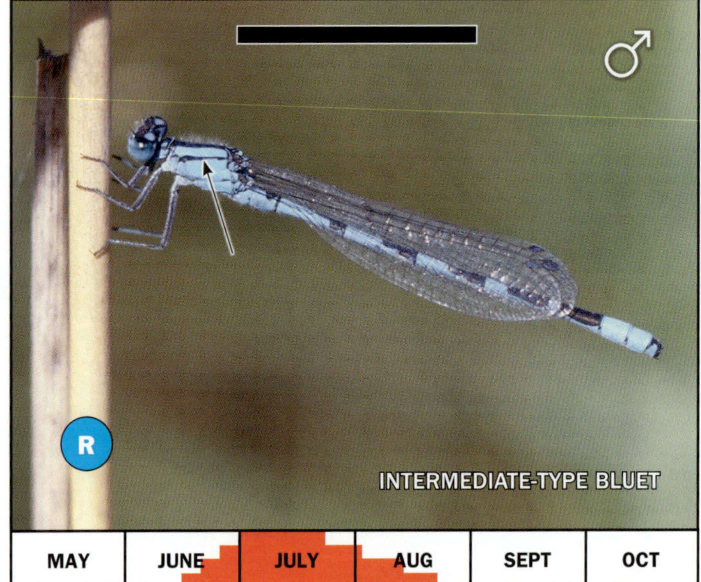

MAY	JUNE	JULY	AUG	SEPT	OCT

Breeding: Alkaline and saline lakes in its western range. Known from wave-swept shores of very large freshwater lakes in our region.

Nature Notes:

An uncommon species in our region but known from several lakes in Minnesota and recently discovered in Wisconsin along the Lake Superior shore.

Flies alertly and low over open beaches and can be hard to catch.

Often lands on sand or other flat surfaces.

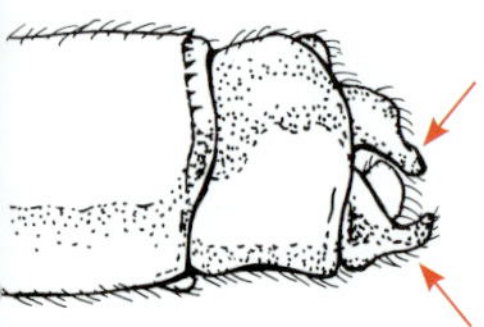

male's abdomen tip; cerci are down-sloping. Paraprocts curve up at tip.

Description: Adults average 1.3 inches long. The male of this enigmatic, intermediate-type bluet has a blue and black thorax with a narrow dark shoulder stripe (↑). Male's abdomen with increasing amounts of black on segments 2 through 7; segments 8 and 9 are all blue. Female's pale areas are blue to yellowish green with her thorax markings similar to the male's. The top of her abdomen is mostly dark, but with segment 8 entirely pale (↑). Eyespots of both sexes large to moderate in size.

Identification Clues: Separate male from other intermediate-type bluets in the hand by looking for the short, downward-angled cerci in side view (see illustration this page). Sometimes confused with male Northern and Boreal bluets but has more black on mid-abdominal segments than those species and cerci shapes are distinctive with a careful look. On females, look for the

The female Alkali Bluet is similar to the Boreal and Northern Bluets, all of which have thin, dark shoulder stripes and can have extensive pale areas on top of abdominal segment 8.

entirely pale segment 8 (but note that female Boreal and Northern Bluets can also have a pale segment 8). Firm identification of females requires seeing mesostigmal plates under a microscope.

Life Cycle & Behavior: Little is known about this species in our region. Flight period appears to be concentrated during mid-summer, with most adults seen here in July. Females oviposit in tandem into masses of filamentous algae and perhaps other vegetation, also into collected mats of vegetation along wave-lapped shorelines. Nymphs overwinter in one of the later instars.

Northern Bluet *Enallagma annexum*

MAY	JUNE	JULY	AUG	SEPT	OCT

Breeding: Well-vegetated shallow ponds, marshes and vernal pools, usually fish-free. Also found in some bogs and fens.

Nature Notes:

Formerly known as *Enallagma cyathigerum.*

While laying eggs, the female may stay underwater for 90 minutes!

Nymphs do best in ponds that lack fish in eastern part of range, not so in West.

Very similar to the Boreal Bluet but appears a bit later in the season and may be less tolerant of acidic conditions.

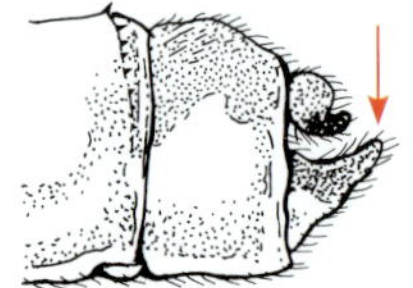

male's abdomen tip; cerci have upturned tips

Description: Adults average 1.3 inches long. Male of this blue-type bluet has a blue and black thorax with a narrow, dark shoulder stripe (↑). Male's abdomen has segments 2 through 5 mostly blue, 6 and 7 mostly black and segments 8 and 9 all blue. Female's pale areas blue or olive brown with thorax markings similar to male. The top of her abdomen is mostly dark, often with a pair of pale spots of variable size on top of segment 8 (↑). Both sexes have large eyespots; the male's nearly touch the eye.

Identification Clues: Male is very similar to the Boreal Bluet and can only be separated from that species by comparing the shapes of cerci (see illustration this page). Tips of the cerci are upturned on Northern, easily seen in side view with a hand lens; tips of the cerci also upturned on Boreal, but turned inward (medially) so not visible in side view. Female can only be

The female Northern Bluet is similar to the Boreal and Alkali Bluets, all of which have thin, dark shoulder stripes and can have extensive pale areas on top of abdominal segment 8.

separated under a microscope from other female bluets with narrow, dark-shoulder stripes and a partially pale abdominal segment 8.

The Vernal Bluet (*Enallagma annexum vernale*) is a closely related subspecies, differing in subtle aspects of the male cerci and perhaps in habitat and habits. Some experts regard it as deserving of full-species status, but populations in our region have many intergrades with Northern and even experts having trouble sorting them out. Because of the great difficulty associated with its identification, the status of the Vernal Bluet in our region is unclear.

Life Cycle & Behavior: Emerges and flies early, reaching peak abundance in June and early July. Females lay eggs into vegetation while in tandem near the surface, or alone into submerged plant stems while males guard nearby. If the female gets caught in the surface tension upon reemerging, the male may "rescue" her by forming tandem and towing her to safety. Nymphs live underwater for about a year and overwinter in one of the later instars.

Boreal Bluet *Enallagma boreale*

MAY	JUNE	JULY	AUG	SEPT	OCT

Breeding: Wide range of still-water habitats: shallow, well-vegetated ponds, marshy lake- shores, prairie sloughs, bogs.

Nature Notes:

The first bluet and one of the earliest damselflies to emerge in our region.

One of the most abundant damselflies in the Far North.

When a nymph loses a leg, it is able to regenerate a new nearly normal one.

Nymphs survive best in ponds that lack fish.

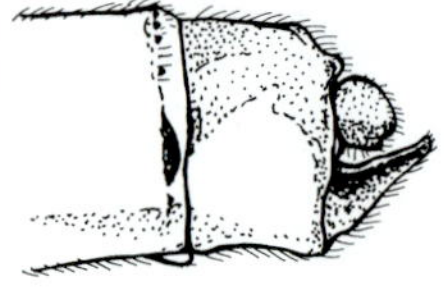

male's abdomen tip

Description: Adults average 1.3 inches long. Male of this blue-type bluet has a blue and black thorax with a dark shoulder stripe narrowing rearward (↑). Male's abdomen has segments 2 through 5 mostly blue, 6 and 7 mostly black and segments 8 and 9 all blue. Female has pale areas blue to olive or tan with thorax markings similar to male. The top of her abdomen is mostly dark with a pair of blue spots of variable size on top of segment 8 (↑) (often fused). Both sexes have large eyespots; the male's nearly touch the eye (↑).

Identification Clues: Very similar to the Northern Bluet, from which it can be separated only in the hand by comparing the shapes of the male's cerci (pages 68-69). Boreal cerci look rounded at the ends in side view because the tips are turned inward (medially); the tips of the cerci are not turned medially on Northern,

Female Boreal Bluets may be blue-form, like this one, or olive-tan. This species is similar to the Northern and Alkali Bluets, all of which have thin dark shoulder stripes and can have extensive pale areas on top of abdominal segment 8.

so they are easily seen as upturned in side view. Female can only be separated under a microscope from other bluets with narrow, dark-shoulder stripes and partially pale abdominal segment 8.

Life Cycle & Behavior: Emerges and flies early, reaching peak abundance in June. Females oviposit either in tandem or alone into emergent plant stems at or near the water's surface. Mating and egg-laying takes 1 to 1.5 hours. Nymphs live underwater for about a year and overwinter in one of the later instars.

Marsh Bluet *Enallagma ebrium*

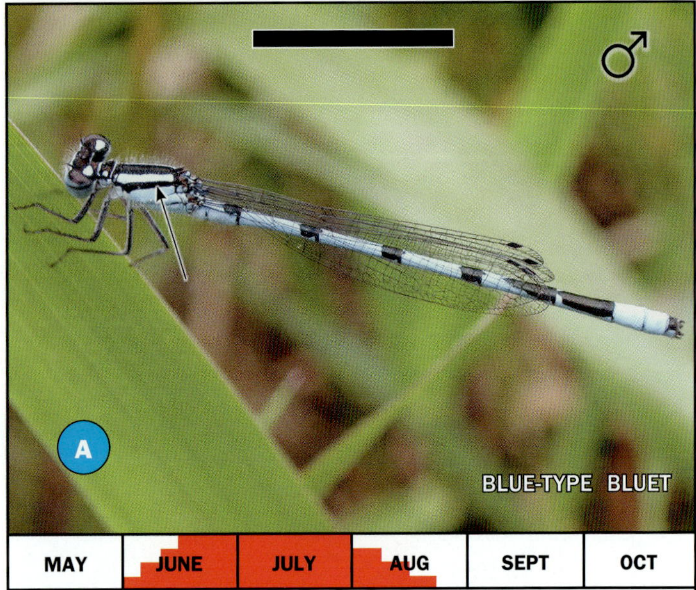

MAY	JUNE	JULY	AUG	SEPT	OCT

Breeding: Lakes, ponds, marshes and slow streams that have marshy borders. Most abundant in waters on calcareous soils.

Nature Notes:

Species name means "drunken," perhaps due to its occasionally erratic flight.

Virtually identical to Hagen's Bluet (except for male's cerci), but does not tolerate acidic conditions as well as that species does.

One female remained underwater laying eggs for five hours!

male's abdomen tip; cerci with 2 arms equal in length.

Description: Adults average 1.1 inches long.

The male of this small, very common, blue-type bluet has a blue and black thorax with a dark shoulder stripe of moderate width (↑). Male's abdomen is mostly blue on segments 2 through 5 with increasing areas of black on 6 and 7; segments 8 and 9 are all blue. Female's thorax is similar to male's, but may be light green, tan or blue where he is blue. Her abdomen is all dark above (↑). Eyespots of both sexes moderate in size.

Identification Clues: Male is most like Hagen's bluet, but differs in having distinctive C-shaped cerci with two arms about equal in length (see illustration this page). Female Marsh and Hagen's Bluets differ from all other bluets by having wide mesostigmal plates and the top of the abdomen entirely black, but can only be reliably separated from each other under a

Note that the top of the female Marsh Bluet's abdomen is entirely dark on all segments. This trait is shared only with the female Hagen's Bluet. While females of all species of American bluets are largely black on top of the abdomen, other species have small pale areas at the bases of some abdominal segments or at the abdomen tip.

microscope. As usual, identity of associated males gives strong clue to identity of females.

Life Cycle & Behavior: Emerges in June over a three-week period and flies through the middle of August. Pairing and egg laying begin in late June and continue through July. Females oviposit alone or in tandem into floating plants, dead or alive, or plants stems below the surface, sometimes becoming fully submerged by a foot or more. Nymphs overwinter in one of the later instars.

Hagen's Bluet *Enallagma hageni*

MAY	JUNE	JULY	AUG	SEPT	OCT

Breeding: Well-vegetated ponds, lakes, marshes and bogs. Tolerant of acidic conditions.

Nature Notes:

One of the most abundant summer damselflies in our region.

Males are almost identical to the Marsh Bluet. Compare terminal appendages to separate them.

Males may pull reemerging females to safety after egg-laying, not altruistically but to mate with them!

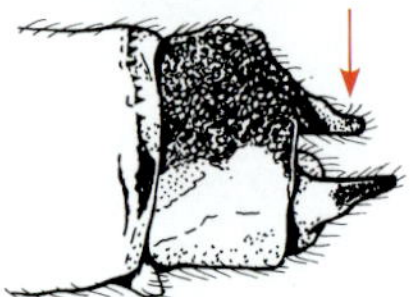

male's abdomen tip; cerci have single horizontal arm

Description: Adults average 1.2 inches long.

The male of this small, very common, blue-type bluet has a blue and black thorax with a dark shoulder stripe of moderate width (↑). Male's abdomen is mostly blue on segments 2 through 5 with increasing areas of black on 6 and 7; segments 8 and 9 are all blue. Female's thorax is similar to male's, but may be light green, tan, or blue where he is blue. The top of her abdomen is entirely dark (↑). Eyespots of both sexes are moderate in size.

Identification Clues: Male is most like the Marsh Bluet, but differs in the shape of its cerci, which have a single horizontal arm (see illustration this page). Female Marsh and Hagen's Bluets differ from all other bluets by having wide mesostigmal plates and the top of the abdomen entirely black, but can only be reliably separated from each other under a

This blue-form female Hagen's Bluet is atypical; most are pale green or tan. Note that the top of her abdomen is entirely dark on all segments. This trait is shared only with the female Marsh Bluet (see female photo caption of that species).

microscope. As usual, identity of associated males gives strong clue to identity of females.

This immature male does not yet show the bright blue of the adults.

Life Cycle & Behavior: Emerges in June and flies into September. Pairing and egg laying begin in late June and continue through July. Females oviposit in tandem into floating plants, dead or alive, or stems just below the surface, without becoming fully submerged. Eggs hatch in about three weeks. Nymphs live for nearly a year and overwinter in one of the later instars.

Stream Bluet *Enallagma exsulans*

MAY	JUNE	JULY	AUG	SEPT	OCT

Breeding: Flowing-water habitats from mid-sized streams to larger rivers; also sheltered lakeshores.

Nature Notes:

Perhaps the most common damselfly along streams and rivers in our region.

Stream Bluets spend over an hour in copulation.

Often found in the company of dancers.

male's abdomen tip; cerci with two arms, lower longer than upper

Description: Adults average 1.3 in ches long. The male of this common, black-type bluet has a blue and black thorax with a dark shoulder stripe wider than the blue stripe (↑). Male's abdomen is mostly black above with segments 3 through 6 having thin blue rings at the bases and segment 9 is all blue (↑). Segment 8 with blue side and dark mark on top narrowing toward apex. Female is similarly marked, but is light green instead of blue on the thorax and the dark shoulder stripe is usually divided in the middle with brown (↑). Female shows abdominal segment 10 and much of segment 9 blue, with dark W-shaped mark at base. Eyespots of both sexes are small and narrow, connected by occipital bar (↑).

Identification Clues: The male is the only black-type bluet in our region with a blue thorax and only abdominal segment 9 entirely

♀

Femaile is quite similar to the male but her thorax is green instead of blue. Note brown center to female's shoulder stripe (inset). Both sexes have an occipital bar connecting their eyespots.

blue. Rearward pointing dark mark on segment 8 often distinctive. Cerci with two arms, lower longer than upper, also distinctive (see illustration on opposite page). Blue-ringed Dancer and Turqouise Bluet males are superficially similar, so look carefully at black-types males along rivers near southern edge of our region. On female look for her light green thorax, wide black shoulder stripes uniquely divided by brown, and the bluish abdomen tip with dark W-shaped mark.

Life Cycle & Behavior: Emerges in June and is abundant by the end of the month. Remains abundant and pairing through July and into August. Some individuals fly through mid September. Males frequently hover over water. Females oviposit alone or in tandem into submerged aquatic plants and may descend well below the surface. Overwinters as late instar nymph.

Rainbow Bluet *Enallagma antennatum*

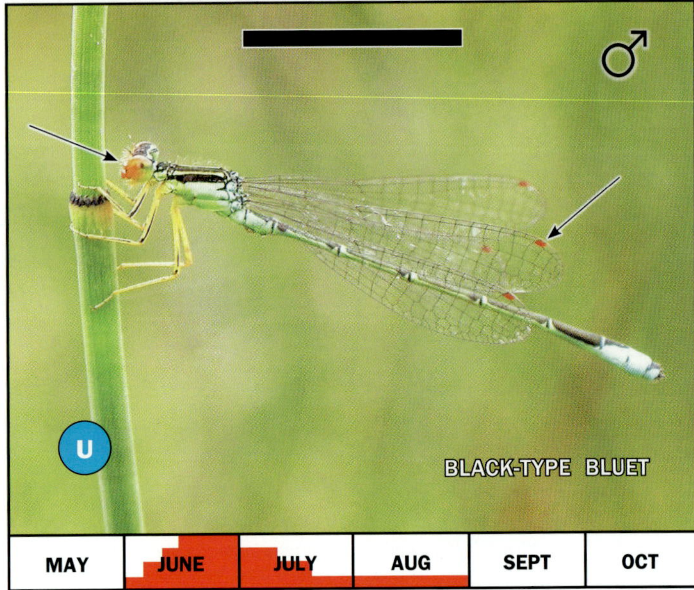

MAY	JUNE	JULY	AUG	SEPT	OCT

Breeding: Quiet streams, usually in open landscapes. Also, springy margins of lakes and ponds especially near outlets.

Nature Notes:

Seeing this strikingly colored bluet for the first time is a notable experience!

Green thorax suggests a forktail more than a bluet, but none of our forktails have a similar color combination.

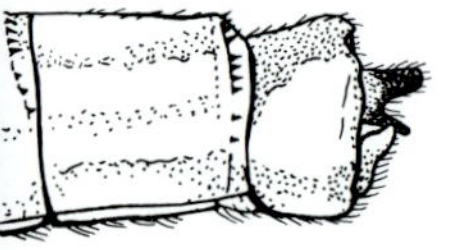

male's abdomen tip

Description: Adults a verage 1.2 inches long. Male is a colorful, black-type bluet with a green thorax, orange face and orange eyes (↑). The dark shoulder stripe is wide and the pale shoulder stripe is narrow and yellow. The top of abdominal segment 9 is entirely blue (↑). His face and front of eyes orange. Female's eyes yellowish brown, sometimes with a trace of orange. Her thorax similar to male's, but greenish yellow (↑) where male's is green. Her abdomen is dark above with an elongate blue or green spot on top of segment 9. Both sexes have narrow eyespots and mostly yellow legs. Wings clear with orange stigmas (↑).

Identification Clues: Male is unmistakable because of his rainbow combination of colors, with orange face easily visible. Female's spindle-shaped blue spot on top of segment 9 is unique, and her yellow legs are often visible in the field.

♀

The thin yellow shoulder stripes, small blue eyespots and orange stigmas are easy to spot on this female Rainbow Bluet.

A male Rainbow showing off his rainbow colors. Note the all-blue segment 9.

Life Cycle & Behavior: Adults emerge from early June to mid July and fly through August. Pairing and egg laying occurs throughout July. While the pair is in tandem, female oviposits into vegetation near water's surface, then she will sometimes descend below to lay eggs into plant stems. Overwinters as a late-instar nymph.

Slender Bluet *Enallagma traviatum westfalli*

MAY	JUNE	JULY	AUG	SEPT	OCT

Breeding: Lakes and larger ponds, usually well-vegetated.

Nature Notes:

Nymphs tolerate fish well.

Extensive blue on head and thorax make both sexes recognizable in the field.

The species has been divided into two subspecies, only one of which, *E. t. westfalli*, occurs in our region.

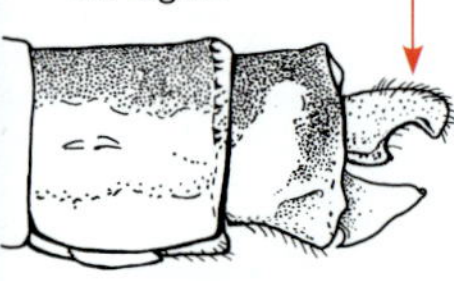

male's abdomen tip; pistol-shaped cercus

Description: Adults average 1.2 inches long. The male of this slender, black-type bluet has synthorax blue with black top and narrow shoulder stripes. Head and prothorax largely blue (↑) with black dividing lines; eyespots large and blue (↑). Male's abdomen black above on segments 3-7, blue on 8 and 9. Segment 2 blue with forward pointing black arrowhead on top. Female similar to male, and usually with pale areas blue (sometimes greenish). Her abdomen mostly dark on top of segments 2-7, blue on segments 9 and 10 (↑), and with a dark longitudinal mark narrowing posteriorly mark on top of mostly blue segment 8.

Identification Clues: Male is unique in showing much blue on top of head (↑) and thorax, combined with mostly black abdomen, and long cerci that are often noticeable in the field. Cercus has long upper arm and large

Female (bottom) is more extensively blue at her abdomen tip than other female pond damsels in our region.

Male shows much blue on top of head.

indent below giving pistol-like look (see illustration on opposite page). The only male bluets with somewhat similarly shaped cerci are River and Azure, but River differs in being an Intermediate-type bluet and Azure differs from Slender in having more blue at abdomen tip with segment 7 mostly blue. Female is more extensively blue at her abdomen tip than other female pond damsels in our region.

Life Cycle & Behavior: Males frequently hover over open water, sometimes while in tandem with the female. Oviposition usually occurs into vegetation at the water's surface with male contact-guarding in sentinel position, but female sometimes descends alone to oviposit into plant stems. Overwinters as nymph in one of the later instars.

Double-striped Bluet *Enallagma basidens*

Breeding: Lakes, ponds (may be temporary), and slow streams with emergent vegetation.

Nature Notes:

Like Great Spreadwing has expanded its range from US Southwest to Northeast during the last century.

Often hovers over open water.

Often abundant at suitable habitat, but local and uncommon here.

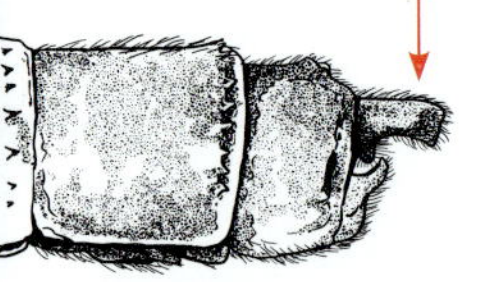

male's abdomen tip; elongate upper arms of cerci extend far past short paraprocts

Description: Adults average 1 inch long. The male of this very small blue-type bluet is blue with black markings including dark top and shoulder stripes on thorax both divided into thin stripes along their entire length (↑). Male's abdomen blue with black apical markings increasing in size from segments 3 through 7. Narrow black line on top of segment 3. Segments 8 and 9 all blue. Eyes entirely blue. Female is light brown on thorax with stripes as in male. Her abdomen all black on top, greenish blue on sides, except segments 9 & 10 which are light blue (↑). Segment 9 has distinctive black pair of longitudinal streaks on top. Eyespots small and narrow with a pale occipital bar between them.

Identification Clues: Tiny with unmistakably doubled thorax stripes. Both sexes are unique among bluets in having the dark shoulder

♀

An exception to the rule, the female is easy to identify because of double thorax stripe, tiny size, and pale blue abdomen tip.

stripe narrowly divided along its entire length, with a thin pale stripe showing in between. Absence of black top on eyes unusual for a male bluet, as is female's abdomen which becomes increasingly blue toward the rear.

Life Cycle & Behavior: Adults emerge primarily in late May and June, and fly into September. Female lays eggs alone or in tandem into floating vegetation or slightly submerged plant stems. Nymphs overwinter in one the last instars.

Orange Bluet *Enallagma signatum*

MAY	JUNE	JULY	AUG	SEPT	OCT

Breeding: Lakes, ponds and slow streams where it often flies low over open water.

Nature Notes:

Reaches greatest abundance in the latter part of summer after most bluets have declined.

Peak activity occurs late in the day, even toward dusk.

Male often flies fearlessly over open water, sometimes far from shore.

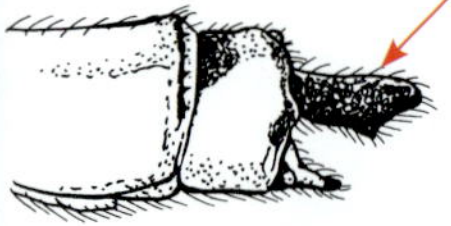

male's abdomen tip; note very long cerci

Description: Adults average 1.3 inches long. The male of this slender, citrus-hued bluet is a study in black and orange. On the thorax, the orange shoulder stripe (↑) is wider than the black one. The abdomen is mostly black with segment 9 entirely orange (↑). Eyes are orange and eyespots small, connected with occipital bar. Female is dull yellow, pale green or blue where male is orange. Her abdomen is mostly dark except for segment 10 and the sides of segment 9, which are pale. Immature males are pale blue (see inset photo).

Identification Clues: Mature males are easily identified by their orange color, including the entire top of abdominal segment 9, and very long cerci (see illustration this page). No other male damselfly in our region has orange at both ends of body. However, immature female Citrine and Lilypad forktails have much orange

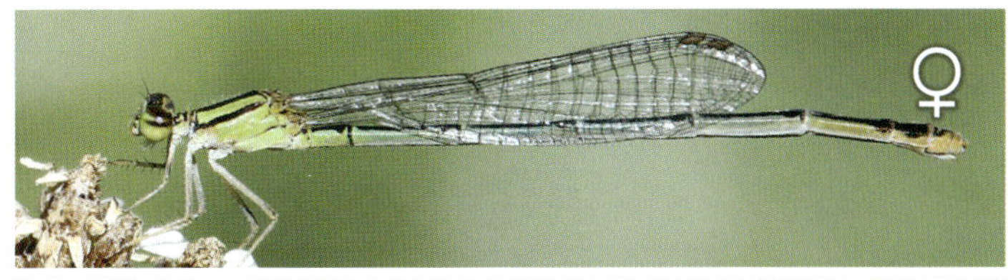

Orange Bluets are not always orange! Most females show yellow or green, but blue-form female Orange Bluets are less common (bottom).

Tiger-stripe coloration of the male is unique among our damselflies.

on thorax and abdomen, including tip. Citrine differs in having orange on abdominal segments 1-5 and lacking dark shoulder stripes; Lilypad differs in having large eyespots and orange on three segments (8-10) at abdomen tip. On the female look for the following: pale shoulder stripe that is at least as wide as the dark stripe, entirely dark top of abdominal segment 8, segment 9 with pale sides and dark top area narrowing to small square at distal end, and pale top of segment 10(↑).

Life Cycle & Behavior: Adults emerge over a long period of at least two months beginning in mid June and flying well into September. Pairing and egg laying occur throughout the summer. Female, while in tandem, lays eggs into floating and emergent plants, with water lilies appearing to be favored. Overwinters as nymph in several late instars.

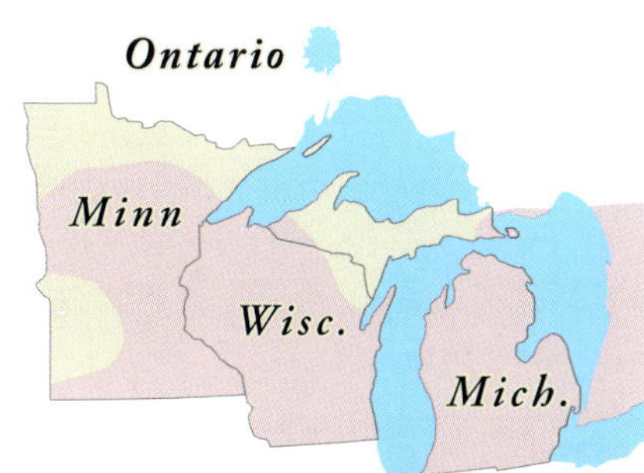

Vesper Bluet *Enallagma vesperum*

MAY	JUNE	JULY	AUG	SEPT	OCT

Breeding: Wide variety of lakes, ponds and slow rivers, usually well vegetated and often with water lilies.

Nature Notes:

Typically flies and mates well into the evening (hence its species name) – bring your flashlight to watch it happen!

Probably more common in our region than thought; infrequently seen because of its crepuscular habits.

Flies low over the water, often landing on lily pads.

male's abdomen tip

Description: Adults average 1.3 inches long. The male of this slender citrus-hued bluet is beautiful and utterly unmistakable. Male thorax is bright yellow (↑) with the dark shoulder stripe very thin to nearly absent (↑); abdomen is mostly black with segment 9 entirely blue (↑). Female similar to male, but her thorax usually greenish yellow and the top of her abdomen dark except for segment 10 and the sides of segment 9, which are pale blue or green. However, her pale colors can be quite variable, ranging from bluish green to yellow.

Identification Clues: Males are easily identified by the combination of yellow thorax and blue abdomen tip. Note the black quadrangle on top of his abdominal segment 10 (↑). His cerci same basic shape as male Orange Bluet, but shorter and wider (compare illustrations on pages 68-69), and his dark shoulder stripe

Females can be yellow, bluish green or greenish yellow, as shown above.

Male Vesper Bluets start adulthood pale blue, gaining the bright yellow thorax only at full maturity. Note male's black quadrangle atop segment 10 (inset).

is thinner (↑). On the female look for the following: a pale shoulder stripe that is very thin or indistinct in middle (↑), entirely dark top of abdominal segment 8, a pale segment 10 and pale sides of segment 9.

Life Cycle & Behavior: Adults begin to emerge in June and have a long flight period that often extends into September. Egg laying occurs in the evening, while in tandem. Eggs are deposited into the leaves of water lilies and perhaps other floating or emergent plants. Overwinters as nymph in several late instars.

Turquoise Bluet *Enallagma divagans*

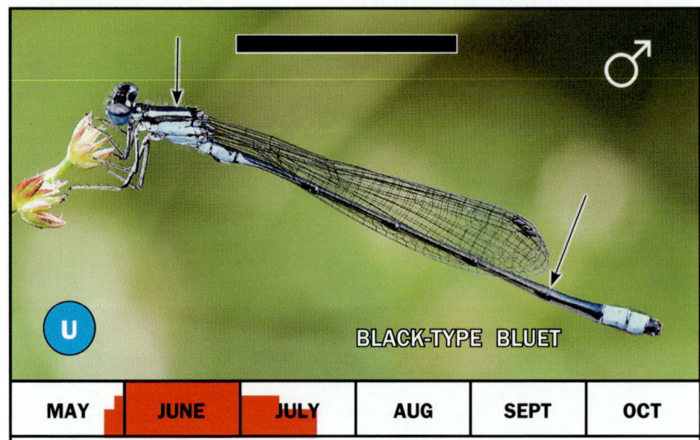

MAY	JUNE	JULY	AUG	SEPT	OCT

Breeding: Slow streams, rivers, and sloughs, and sometimes lake and reservoir shorelines, typically in forested landscapes.

Nature Notes:

Males typically have a dark look and only occasionally show turquoise on the sides of the thorax. They fly slowly and hover frequently, sometimes just inches above water.

Known in our region only from southeastern Michigan where it is not common.

Early season species that emerges in late May.

Females oviposit in tandem into vegetation near the water line, but sometimes descend well below to oviposit alone while the male guards from above.

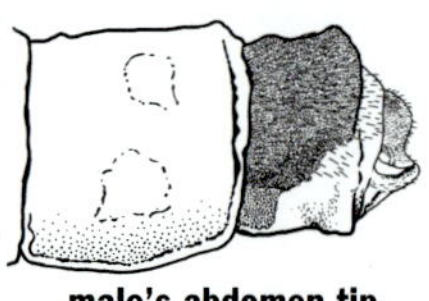

male's abdomen tip

Description: Adults average 1.2 inches long. A small, black-type bluet of slow-moving waters. Male's thorax has blue and black shoulder stripes about equal in width (↑). His abdomen is black on segments 3-7 (↑) and 10, without pale rings on the middle segments. Segments 8 and 9 all blue. Female is similar, but her dark shoulder stripe often with some brown in middle. Her abdomen tip with segment 10 and most of segment 9 blue; segment 8 with some blue. Both sexes have eyespots narrow with an occipital bar.

Identification Clues: Slender and Prairie have larger eyespots and mid-abdominal pale rings; Skimming has more black laterally on segments 2, 8, and 9; See also Stream and Lilypad accounts.

Forktails Genus *Ischnura*

Forktails are small, strikingly patterned damselflies, similar in general appearance to the American and Eurasian bluets. They are set apart from these genera by the variation in color of females and the small forked structure at the tip of the abdomen (segment 10) of males. The famous Canadian odonatist E. M. Walker aptly called this forked projection a "turret-like eminence, which is more or less distinctly bi-lobed." Females come in a wide array of colors for two reasons. First, most of our species have a male-like form (homeochromatic). Further, females change color as they age by gaining a grayish-white or bluish pruinescence that gradually obscures much of the underlying color pattern. The scientific name *Ischnura* means "slender-tailed," referring to their thin abdomen.

Forktails additionally have some unusual life history features that set them apart from other groups. Most of our species have long flight periods, beginning to emerge in the spring and continuing to emerge and fly throughout the summer. Newly emerged forktails reach reproductive maturity more rapidly than other damselflies, sometimes in just one day. Females of some species may mate only once (unusual for damselflies) and they usually oviposit alone. Forktails are intermittent fliers that hold close to the emergent vegetation along the shore of their breeding habitat, yet, the Citrine Forktail manages to be a very effective and wide-ranging disperser.

This abundant genus of pond damsels is found in still-water habitats throughout the world. There are 14 species in North America, of which six are residents of our region (or at its periphery). The Eastern Forktail is one of our most abundant and widespread damselflies, while the other species are less common. Males are readily identified in the field by their distinctive color patterns, except for Eastern and Western forktails, which should be examined in hand at the western edge of our region. Confirm identifications by viewing the terminal appendages with a hand lens. Younger females can often be identified in the field or in the hand by their distinctive color patterns, but the pruinescence on older females largely obscures these patterns. Often, enough of the pattern can still be seen to identify them. When in doubt about the identity of females, it can be useful to identify nearby males.

Lilypad Forktail *Ischnura kellicotti*

MAY	JUNE	JULY	AUG	SEPT	OCT

Breeding: Lakes and ponds of various sizes having extensive coverings of water lilies (both *Nuphar* and *Nymphaea*).

Nature Notes:

Virtually all of the adult's activities are associated with water lilies, with which the species has an obligatory relationship.

Females are thought to mate only once.

Flights between perches are low and quick.

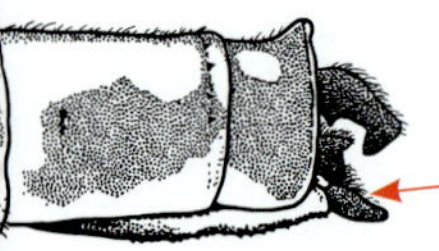

male's abdomen tip; note forked paraprocts

Description: Adults average 1.1 inches long.

Heavily marked forktail invariably associated with water lilies. Male has blue thorax with wide black shoulder and median stripes. His abdomen mostly black with a blue ring at the distal end of segment 2 (↑), and blue at the abdomen tip on top of segments 8 and 9, and at the distal end of segment 7. Female is patterned much like male, orange when immature and becoming grayish white pruinose, obscuring underlying markings, as she ages. Older females can appear almost white. Both sexes have large eyespots.

Identification Clues: Male is easy to recognize, being heavily marked with black, having a blue ring on segment 2, and with blue at the abdomen tip only on the top half of the segments. No other forktail or black-type bluet is so marked. Male Skimming Bluet has wavy blue line, not ring, on segment 2. Examine male's

Female is patterned much like male, orange when immature (top) and becoming grayish white pruinose, obscuring underlying markings, as she ages (bottom). Older females can appear almost white.

forked paraprocts in side view to confirm his identity (see illustration on opposite page). Female's heavy black markings unlike any other female damselfly in our region (but see cautions on Orange Bluet page). As female gains pruinosity, identification is less certain, but the eventual almost all white appearance of her thorax a good clue. Tentatively identify questionable females by the males they are associating with.

Life Cycle & Behavior: Adults are present over a long period emerging June and flying into August. Both sexes are usually found perched on water lilies with their abdomens slightly down-turned to contact the plant. Females oviposit alone into lily pads, taking up to 20 minutes to do so. Nymphs crawl about on the underside of lily pads seeking prey and crawl on top when ready to emerge. Overwinters in several mid and late nymphal instars.

Plains Forktail *Ischnura damula*

Breeding: Wide variety of well-vegetated, still-water habitats: lakes, ponds, ditches, warm/hot springs, and sluggish streams.

Nature Notes:

Has a highly disjunct distribution in western Canada, with isolated populations occurring only at hot springs.

Scientific name means "small deer," likely referring to the forked projection on top of abdominal segment 10 being reminiscent of a young buck's antlers.

male's abdomen tip; abruptly upturned paraprocts

Description: Adults average 1.1 inches long. Small, attractive forktail with dotted thorax. Male's pale shoulder stripes interrupted to form four widely spaced blue dots on his black thorax top (↑). Sides of his thorax blue. Abdomen is black except for segments 8 & 9 which are blue with black rectangles on the sides. Forked projection on his abdomen tip not pronounced. His head is dark with eyes black above, green below, and small, blue eyespots. Females show several colors and patterns. Typical-forms can be blue, greenish brown, or orange-pink on eye spots and thorax with thin, black top and shoulder stripes when young. With age her thorax darkens to brown and eventually gains pruinosity. Less-common male-like female has thorax markings like male. Her abdomen black above with considerable blue on segment 8 and often partly blue segment 9 regardless of age or form. In all cases females have a prominent, nipple-like tubercle on each side of the prothorax.

Note attractive combination of light pink and blue on the typical-form female above. Blue-form females are uncommon.

Identification Clues: Plains male is the only forktail in our region having four blue dots on top of black thorax. Confirm in the hand by diagnostic shapes of his abruptly upturned paraprocts seen in side view (see illustration on opposite page). Identify male-like females by spotted thorax, and typical females by the prominent tubercles on the prothorax seen with a hand lens. No need to be confused by the wide array of her potential colors depending on form and stage of development. As usual with questionable females, look at males they are associating with.

Life Cycle & Behavior: Fairly long flight period that extends into September; adults begin to emerge in May and continue into summer. Adults mature quickly after emergence. May perch with wings slightly spread, especially right after landing. Perches in thickly vegetated areas near water. Female lays eggs alone into aquatic vegetation at the water surface. Little is known about the life history of this species. Overwinters as nymph.

Western Forktail *Ischnura perparva*

Breeding: Lakes, ponds, and sluggish streams with grasses and sedges along shore; often at mud-bottomed, alkaline habitats.

Nature Notes:

Female brushes off unwanted suitors by fluttering her wings and curling her abdomen tip downward.

Not afraid to fly low and straight over the water, unusual for a forktail.

May not range East far enough to reach western periphery of our region, but close enough that folks in that area should examine males closely.

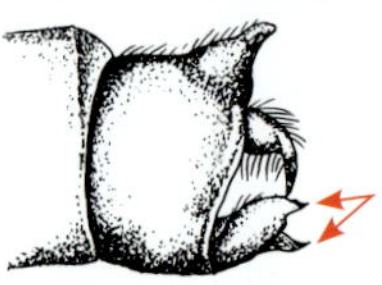

Upper and lower forks of paraproct about equal size. Compare to Eastern Forktail.

Description: Adults average 1.1 inches long. A small, dark and strikingly marked species. Male's thorax is green with thick black top and shoulder stripes. Abdomen is black above from base through segment 7; segments 8 & 9 blue with black rectangles on the sides (↑). His head is dark above with lower eyes green, and small, bluish green eyespots; face green. Younger female's thorax striped like male, but orange which extends onto base of abdomen and orange also appears in various amounts at tip. Rest of her abdomen black on top. Females turn entirely bluish or light gray pruinose with age (just a few days), but the complete shoulder stripes may remain visible. Uncommon male-like female probably also becomes pruinose with age.

Identification Clues: Rectangular dark marks on sides of blue abdomen tip of males, in conjunction with shoulder stripes (not dots)

Scientific name means "very small," which it is, but a few other forktails average even smaller.

on thorax, narrow possibilities to Western and Eastern Forktails. Separate these two by shapes of forked paraprocts (primarily) and cerci (secondarily) in side view (see illustration on opposite page). Western differs from Eastern in having longer paraprocts, the forks of which are about equal size. Throughout most of our region, Eastern is the only one of these two species that will be found, but when near the western periphery of the region, be particularly careful about examining males in hand. Females are very much like female Eastern; can be identified with certainty only by examining the mesostigmal plates under a microscope.

Life Cycle & Behavior: Long flight period that extends into October; adults begin to emerge in May and continue through summer. Adults mature in just a few days after emergence. Female mates only once and lays eggs alone into wide variety of vegetation at the water surface. Females frequently outnumber males at water. Overwinters as nymph.

Eastern Forktail *Ischnura verticalis*

Breeding: Wide array of permanent still waters including ponds, lake bays, marshes and slow streams.

Nature Notes:

One of the most widely distributed and abundant damselflies in our region.

Females often flutter their wings at males, but this seems to be more of a threat than a courtship display.

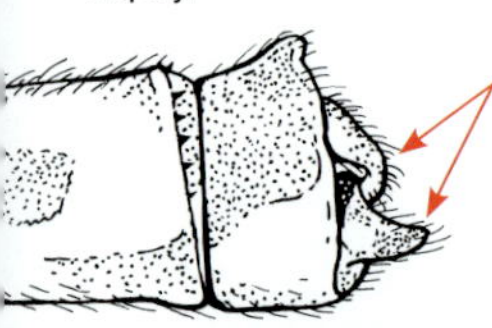

male's abdomen tip; Lower arm of paraproct much longer than upper. Compare to Western Forktail.

Description: Adults average 1.2 inches long.

A very common, strikingly marked species. Male's thorax isgreen with thick black top and shoulder stripes (pale shoulder stripe sometimes interrupted). Abdomen is black above from base through segment 7; segments 8 and 9 blue with black rectangles on the sides. His head is dark above with lower eyes green and with small, bright green eyespots. Younger female's rear of head, most of thorax and base of abdomen are all orange. She has a black, thin shoulder stripe, black mid-dorsal stripe and most of her abdomen is black above. Females turn entirely bluish pruinose with age, but the complete shoulder stripes may remain visible. Uncommon male-like female probably also becomes pruinose with age.

Identification Clues: Male's striped thorax in conjunction with rectangular dark spots (↑)

Female Eastern Forktails come in two very different color varieties. Immatures are orange (bottom) and turn bluish-gray pruinose with age (top). Occasionally, the male Eastern Forktail shows a divided shoulder stripe much like the Fragile Forktail (inset photo right).

on the sides of segments 8 and 9 at the blue abdomen tip are definitive throughout most of our region. Western Forktail, possibly at the western periphery of our region, is our only other species so marked. The two species differ in Eastern having shorter paraprocts, the two tips of which are unequal in size (see illustration on opposite page). Females have complete pale and dark shoulder stripes and a small vulvar spine (the only other forktail that has this combination in our region is the Western Forktail) .

Life Cycle & Behavior: Adults begin to emerge in late May and continue to emerge throughout the summer. Some individuals may still be flying in October. Young adults mature quickly (males gain full color in one day) and do not fly far from the breeding site. Females may only mate once, storing that sperm for all future egg fertilization. Females oviposit alone (usually) into vegetation near water's surface. Nymphs overwinter in various instars.

Fragile Forktail *Ischnura posita*

MAY	JUNE	JULY	AUG	SEPT	OCT

Breeding: Slow streams, ponds, and marshes that are shaded, heavily vegetated and sometimes spring-fed.

Nature Notes:

One of our most shade-tolerant damselflies.

Species name means "positive," probably alluding to the exclamation point on the thorax!

Despite tiny size and serene demeanor, female is known to prey on other damselflies.

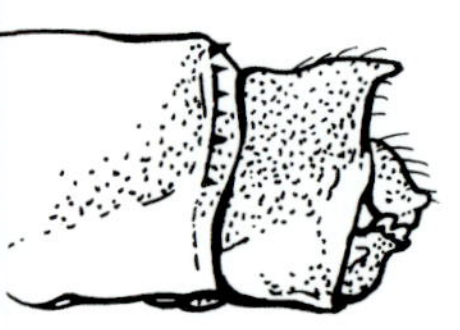

male's abdomen tip

Description: Adults average 1 inch long. Very small, dark species. Male's thorax and abdomen are black above with a pale green shoulder stripe interrupted to form a dot and short line like an exclamation point (!) (↑). Very thin, pale rings around the base of each abdominal segment. Female is very similar, but may have pale blue (↑) or green on her thorax. Female is never orange. She turns grayish as she ages, but the interrupted shoulder stripe usually remains visible.

Identification Clues: Both sexes are readily identified in the field by pale shoulder stripe divided like an exclamation point in conjunction with entirely dark abdomen. Male's dark abdomen top (↑) is unique; males of all other pond damsels in our region have at least one entirely pale segment on top of their abdomen (pale segment may be blue, green, red, orange or yellow).

Female Fragile Forktails are patterned much like the males but they may have blue on the thorax shading to gray with age.

Life Cycle & Behavior: Adults have a long flight period from late May well into September. Females oviposit alone into stems of emergent vegetation at or slightly above water's surface. Known to roost at night in same vegetation where found during the day, but higher off the ground. Probably overwinters in one of the final nymphal instars.

Note the pair of divided shoulder stripes like tiny exclamation points. The Fragile Forktail is the only damselfly in our region to ALWAYS show this mark. Eastern Forktails sometimes do but they never are entirely dark on top of the abdomen.

Citrine Forktail *Ischnura hastata*

Breeding: Lake and pond margins, & permanent areas of spring seepage; often thickly overgrown with grasses, sedges, spike rush or club-rushes.

Nature Notes:

Called "exquisite little creatures" by pioneering odonatists Needham and Heywood.

Like the Sphagnum Sprite, a rare, tiny damselfly of our region.

Small enough to be hard to see in their densely-vegetated habitats; but bright yellow males sometimes easy to see against right background.

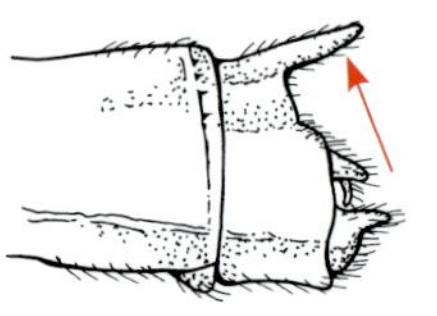

male's abdomen tip; note male's two-pronged fork on top of segment 10.

Description: Adults average 0.9 inches long. Tiny yellow and black damselfly. Male has a dark top of the head with tiny blue-green eyespots. Thorax is dark above and yellowish-green below with thin pale shoulder stripes. Abdomen is largely yellowish-orange with regularly spaced dark marks. When viewed from above, the tip of the abdomen shows a very prominent two-pronged fork (↑). Young females are orange with dark marks on top of the thorax and on top of abdominal segments 6 through 9. Females darken with age, becoming increasingly bluish-gray with some pruinosity.

Identification Clues: The male is easily recognized in the field by his small size, mostly yellow abdomen and prominent, slender fork on top of segment 10. Orange stigma in forewing of male is unique in being well separated from the wing margin (↑), Look for the very thin or

This specimen shows the heavy bluish-gray pruinosity of an older female Citrine Forktail. When at this stage, they are difficult to separate from other female forktails.

Note the very thin, dark shoulder stripe and overall orange color of this young female.

absent shoulder stripe (↑) on young (orange) females; older females are difficult to separate from other pruinose forktails although the lack of a dark shoulder stripe may remain evident.

Life Cycle & Behavior: Adult flight period is fairly long, from early June into September. Females oviposit alone into stems of emergent vegetation. Females mate just once, storing sperm for future egg fertilization. Parthenogenetic populations (lacking males) occur in Azores. Excellent disperser that is carried about for great distances by high winds. Populations in our region may not breed successfully here; probably arrive here from the south. Probably overwinters in one of the final nympal instars.

Western Red Damsel *Amphiagrion abbreviatum*

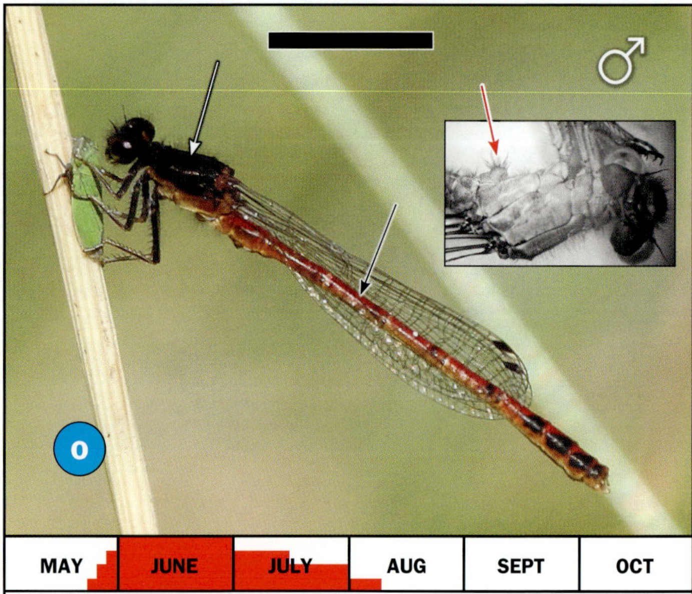

Breeding: Gently flowing spring-fed streams, spring ponds or spring upwellings. Also at margins of bogs, ponds or ditches.

Nature Notes:

Its need for spring upwellings in our region means it may disappear from areas where the water table has been lowered.

Areas of preferred habitat are often small, but capable of producing many individuals.

Flies close to the ground in vegetation, avoiding open water.

male's abdomen tip; thin cerci slope downward.

Description: Adults average one inch long.

Small, stocky, two-tone species. Male is mostly red (↑) with black areas (↑) that increase in size with age on head, top of the thorax and on the last three or four abdominal segments. Female is similar but paler, with little or no black on her head or thorax. Her thorax and abdomen are mostly brownish orange with reddish highlights on the abdomen (↑) and dark spots on tops of segments 7-9. Her head and eyes are reddish brown without eyespots. Wings clear. Both sexes with a prominent, hairy "bump" on underside of thorax (↑). Female has strong vulvar spine.

Identification Clues: Male is our only two-tone, red-and-dark damselfly that lacks eyespots. His downward sloping cerci (seen in the hand) are diagnostic (see illustration this page). On female look for the lack of eyespots, little or no black areas on thorax, and some dark areas at the

The female Western Red Damsel's completely reddish-orange body with no black areas or eyespots separates her from all other species. Note the mites on the underside of her thorax.

Red Damsels, Genus *Amphiagrion*

This beautiful little genus contains only two very closely related species in North America: the Eastern Red Damsel and Western Red Damsel. Interestingly, populations in the Upper Midwest are intermediate in some characteristics between the two species and debate continues about how many species should be named. Specimens in our region have long been considered to be most like the eastern species, but the most recent morphologic and genetic evidence now suggests they are the western species and will be referred to as such until new study results shed further light on the situation. Nymphs of this species and the Aurora Damsel are distinctive in having sharply angled rear corners of the head. Genus name means "both Agrion" (a word for damselfly), probably referring to the two species in the genus.

tip of the abdomen; other female pond damsels that show orange have eyespots and black areas on the thorax. "Bump" on underside of thorax of both sexes also unique (inset photo on opposite page).

Life Cycle & Behavior: Adults emerge in late May or June and fly though July. Females oviposit in tandem, often with male in "sentinel" position, into wide variety of floating vegetation. Probably overwinters in one of the last few nymphal instars. Nymphs have been found in seepage areas with very little water.

Aurora Damsel *Chromagrion conditum*

Breeding: Shady, spring-fed streams, especially pools and slow backwaters; sometimes clean ponds and rivers.

Nature Notes:

Perched males hold their wing partially spread apart, though not quite as widely apart as spreadwings.

Populations in our region are usually quite local, with low numbers seen.

Adults stay close to shelter and spend little time on the wing.

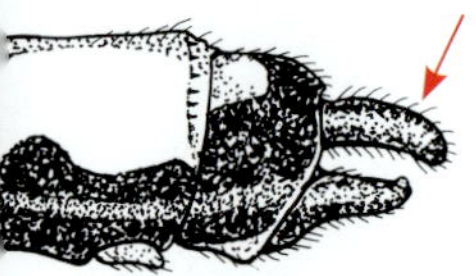

male's abdomen tip; long curved cerci are unmistakable

Description: Adults average 1.4 inches long. A large and slender, strikingly colored pond damsel. Male's thorax is distinctive; unstriped black above with wavy edges (↑), blue sides and a unique yellow blotch (↑) on the lower side. Abdomen is mostly black with thin, blue rings on middle segments and blue patches on top near the tip. Female is similar including black on top of thorax with wavy edging and a yellow side blotch that may be pale. Her thorax may be blue but is usually pale brown, yellow, or gray where his is blue. Wings clear. No eyespots.

Identification Clues: The only damselfly in our region with a thorax having a large black area with wavy edges on top, no shoulder stripes and a yellow side blotch on both sexes. Long, curved shape of male's cerci is unmistakable (see illustration). All American and Eurasian bluets in our region have obvious eyespots.

You can just see the characteristic yellow blotch on the lower thorax of this immature male Aurora Damsel.

Aurora Damsel, Genus *Chromagrion*

The Aurora Damsel is the only species in its genus, and occurs only in eastern North America. It is unlike other members of the pond damsel family in the way it holds its wings, aspects of wing venation, color pattern of the thorax, and in the long terminal appendages of the males. Nymphs have distinctively angled rear corners of the head, as do the red damsels. Genus name means "colored Agrion (a word for damselfly)," referring to the distinctive blue and yellow hues on the thorax.

Life Cycle & Behavior: Adults emerge in June and fly into August. Emergence perches are plant stems a few inches above water and often further from shore than other species. The male seizes the female while perched and quickly forms tandem position. Egg-laying occurs while in tandem, with the male guarding in the sentinel position. Female inserts eggs in or on plant stems just below water's surface. Hatching occurs about three weeks later. Overwinters in late nymphal instars.

Sprites
Genus *Nehalennia*

The sprites are tiny, dainty damselflies that are easy to identify once you find their haunts. Their small size, unstriped metallic-green thorax and lack of eyespots sets them apart at a glance from all other damselflies. Sprites typically fly low among plants and are reluctant to fly over open water. For this reason and because they are so small, they are inconspicuous and require a bit of patience to observe. Once you find them, they are fun to watch and will often allow you to approach quite closely. These fragile insects are both beautiful and interesting. The name *Nehalennia*, appropriately enough, is after a river goddess of the Rhein. I've noticed that when shown their first sprite, people often respond by saying, "How cute!" or "Isn't it adorable!"

Six species of sprites are known world wide, five from the Americas and just two species are found in our region. Sedge Sprites are very common and widespread in a variety of habitats, whereas Sphagnum Sprites are rare and limited to the sphagnum moss margins around some bog ponds. Identifying the two species requires determining the sex first, which can be a challenge with such a dainty subject! Males can be identified in the field based on the pattern of blue on abdominal segments 8 to 10. Pattern of blue at the abdomen tip is helpful to identify females as well, but they can also be identified in the hand by the shape of the rear edge of the prothorax (see below).

Females in this genus oviposit in tandem in floating bits of dead rushes or other vegetation, or into moist sphagnum moss. The female keeps her body horizontal when egg laying as the contact-guarding male supports himself at a 45 degree angle by holding her prothorax with his claspers.

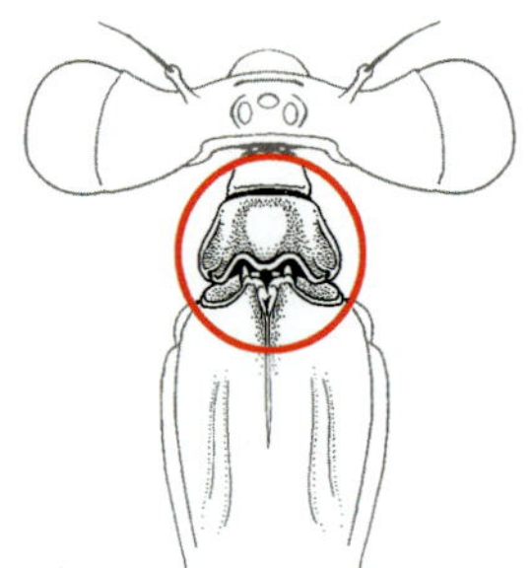

Female Sedge Sprite prothorax: note three-lobed hind margin.

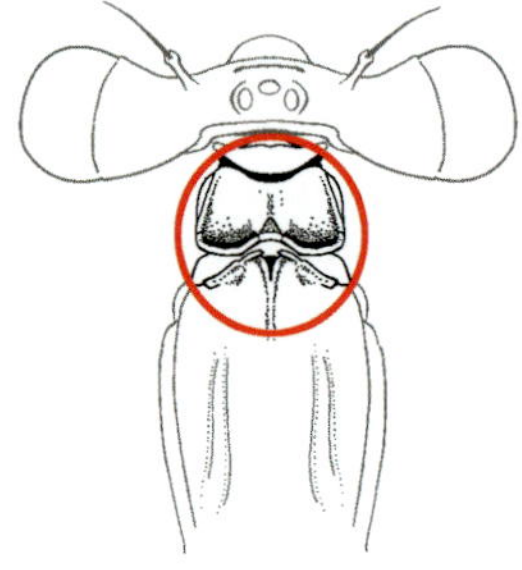

Female Sphagnum Sprite prothorax: note two-lobed hind margin.

Sphagnum Sprite *Nehalennia gracilis*

Breeding: Only in wet sphagnum fringes surrounding quaking bog ponds.

Description: Adults average 0.9 inches long.

Eyes blue, face pale blue. No eyespots, but a thin blue bar is usually present along the back of the head. The thorax and abdomen are unstriped, metallic green on the top half, light blue below. Females and tenerals are yellowish below. Abdominal segments 9 and 10 of male are entirely pale blue (↑) and segment 8 is mostly blue with a small dark area at base. Female lacks vulvar spine.

Nature Notes:

This eastern species is evidently rare in our region, but has hardly been looked for here in the right habitat. Might it be more common here than we think?

One of our daintiest damselflies. Species name comes from its slender abdomen.

Identification Clues: Segments 9 and 10 of male are all blue whereas male Sedge Sprite has small dark areas at the base of both segments. Rear edge of prothorax of female has two wide lobes while female Sedge Sprite has three distinct lobes (see illustration on pg. 128).

Life Cycle & Behavior: Adults emerge in June and fly into late August. Little is known about the life history of this species.

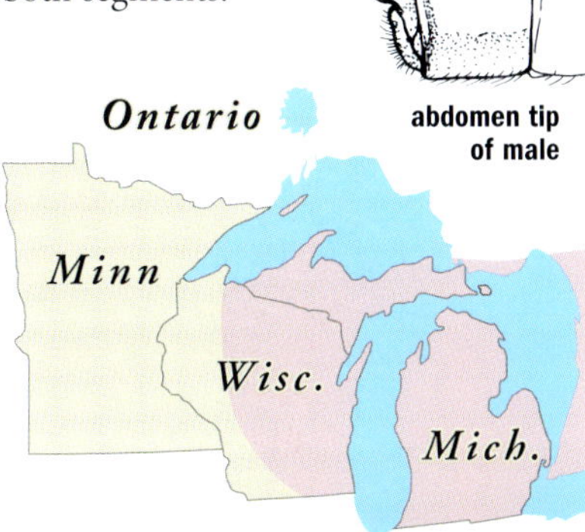

abdomen tip of male

Sedge Sprite *Nehalennia irene*

Breeding: Bogs, sedge fens, marshes, ponds, slow streams, some lakes and vernal pools.

Nature Notes:

One of the most abundant damselflies in our region.

Look for it flying low amongst the dense vegetation surrounding almost any still-water wetland that supports emergent plants

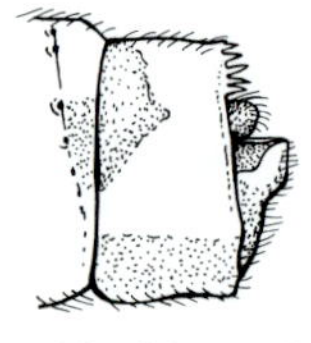

male's abdomen tip

Description: Adults average 1 inch long. Eyes blue, face pale blue. No eyespots, but a thin blue bar is usually present along the back of the head. Male's thorax and abdomen are unstriped, metallic green on the top half, light blue below. His abdominal segments 9 and 10 are pale blue with dark areas at bases of the segments (↑). Females and tenerals are yellowish below. Female has less blue on segment 8 than male, and her vulvar spine lacking or tiny.

Identification Clues: Sprites can only be confused with each other due to their unique combination of tiny size, unstriped, metallic-green thorax, and lack of eyespots. Correct sexing is essential as both species have male-like forms. Segments 9 and 10 of male Sedge is blue with dark areas at the base of both segments. Male Sphagnum has segments 9 and 10 entirely blue. Rear edge of prothorax of female has

Female Sedge Sprites are very similar in color to males but she lacks the blue markings of segments 8 and 9.

Tiny size, metallic-green thorax and blue eyes immediately indicate a sprite.

Female laying eggs as the male contact-guards her.

three distinct lobes while female Sphagnum has only two wide lobes (see illustration on pg. 128). Female Sedge has less pale coloration at abdomen tip than female Sphagnum, but female Sedge sometimes has a male-like form very similar in color and pattern to female Sphagnum. In hand examination may be necessary.

Life Cycle & Behavior: Adults emerge in late May or June and often fly throughout the summer. Female lays eggs into floating bits of dead rushes or other plant material, or into moist sphagnum moss. Overwinters in one of the final two nymphal instars and emerges the following spring.

Dancers
Genus *Argia*

The Dancers are a colorful group of damselflies of medium build and moderate to large size. Males are predominantly blue or purple and females are olive or brown. Males may lose their bright colors when they get cold or are in tandem, but regain them when those conditions change. They are known for their bouncy flight (hence the common name) and their tendency to alight on bare, sunny places such as roads, paths and shoreline rocks, though some species do perch on vegetation. Dancers are alert, tend not to rest long in one spot and may be hard to approach. Our northern species are usually found along flowing waters or the wave-beaten shores of large lakes. Males arrive at breeding sites before females. They fly aggressively at other males, but without physical contact and with little clear defense of territories. Females deposit eggs into wet wood or in mats of floating plant material, often at communal sites. Females usually oviposit in tandem, with males alertly guarding in the "sentinel" position, but they may also be unattended. Powdered dancers oviposit well underwater, the male often remaining in tandem and becoming completely submerged. Nymphs live in gently moving water either under rocks in rapids or in mud and detritus in quieter areas. The nymphs are easy to recognize with their short, stocky bodies and wide caudal lamellae. The most likely meaning for the name *Argia* is "lazy," but no one seems to known how this applies to the group. *Argia* may have been intended to mean "bright," from the Greek *argos*, referring to the coloration of the males.

Although some dancers are similar in size and coloration to Eurasian and American bluets, they are easily separated from these groups by a number of behaviors and physical characteristics. One distinctive habit of dancers is to perch with wings held high, slightly above the abdomen, and not along side touching it, as do bluets. Dennis Paulson has pointed out another remarkable perching difference between the groups: when perched on vegetation overhanging water, dancers tend to face the stream whereas bluets tend to face the shore. Among the physical differences between the groups are the long spurs on their lower legs (tibiae) of dancers that are twice as long as the spaces between them. Bluets have tibial spurs only about as long as the spaces between them. The dark shoulder stripes of dancers are either notched or forked near the upper ends, or are distinctly narrowed near the middle. In contrast, the dark shoulder stripes of bluets are never forked and are either the same width throughout or are gradually narrowed toward the upper end. The terminal appendages of male dancers have cerci shorter than their paraprocts; the reverse is often true of bluets. Dark markings of the abdomens of females are often on the sides of segments of dancers versus on the top of segments of bluets. Female dancers lack a vulvar spine on abdominal segment 8; American bluet females (genus *Enallagma*) have noticeable vulvar spines.

Males of our species are easily identified in the field, or in the hand, by their colors, patterns, and the shapes of their terminal appendages seen in side view. Females can usually be identified in the hand. This is a Neotropical genus of about 113 species, most living in Central and South America. Currently, 32 species are recognized in North America, with six occurring in our three-state region.

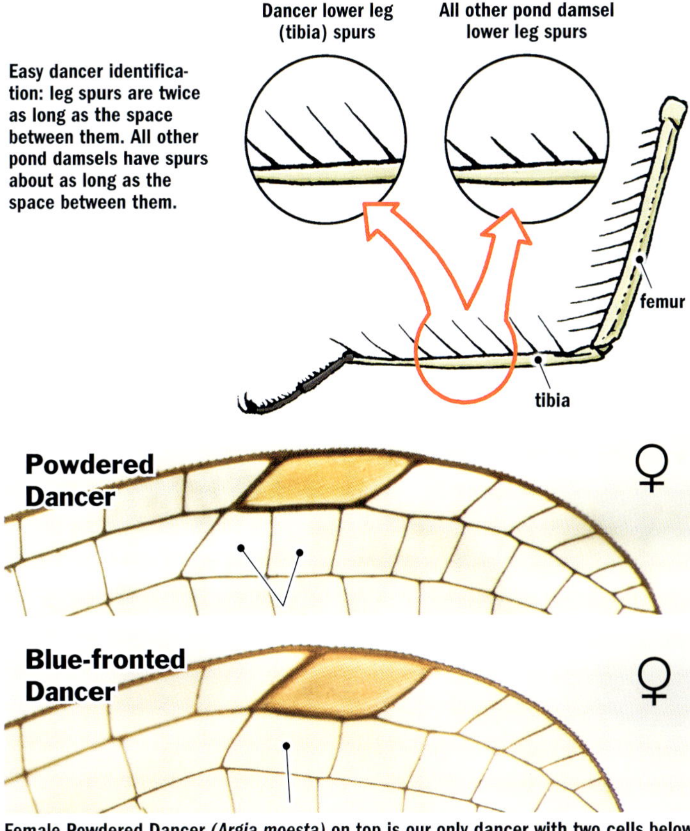

Female Powdered Dancer *(Argia moesta)* on top is our only dancer with two cells below the stigma of the wing, separated in the middle by a crossvein (both sexes). Female Blue-fronted *(Argia apicalis)* has only one cell in wing directly below the stigma.

Blue-fronted Dancer *Argia apicalis*

Breeding: Wide variety of rivers and streams with gentle current and sand or mud bottoms; occasionally lakes and ponds.

Nature Notes:

Flies with a "dancing" movement over streamside vegetation.

Male's bright blue thorax is easily seen when he is perched on bare ground.

Often perches on bare ground, horizontally when on vegetation.

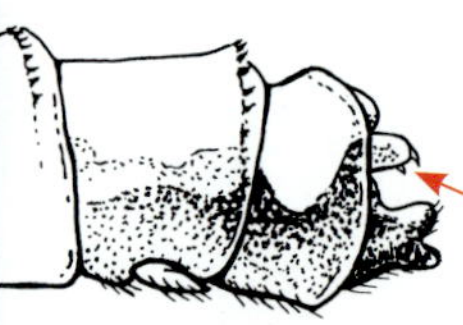

male's abdomen tip; cercus has tiny down-pointing teeth

Description: Adults average 1.5 inches long. A brightly colored and slender species. Males show bright blue on the thorax (↑) and on top of last three abdominal segments while the rest of abdomen is dark. Dark shoulder stripes are hairline thin. His eyes and face blue. Female is either brown on thorax or blue like the male, with dark shoulder stripes hairline thin or absent. Her abdomen is dark on top with a narrow pale stripe on some segments and a dark side stripes on segment 9 (↑). Wings of both sexes clear.

Identification Clues: Bright blue thorax of male with only a hairline shoulder stripe is unmistakable. Shape of his cercus is also distinctive. Brown-form female is very similar to female Powdered Dancer, both having very thin shoulder stripes. Female Blue-fronted has only one cell in wing directly below the stigma (↑) whereas

Brown-form female Blue-fronted Dancers blend well with the rocky or sandy surfaces they normally land on. Inset: Note only one cell below wing stigma.

This blue-form female's thorax is colored like the male's.

female Powdered Dancer usually has two cells there (with a crossvein below the middle of the stigma). Female Blue-fronted usually more heavily striped than female Powdered on top of segments 8 and 9, and on the sides of segment 9, but much variation in this, so focus on wing-cell difference and identity of males they are associating with.

Life Cycle & Behavior: Adults have a long flight period, beginning to emerge in mid June and flying into September. Individual adults live about a month at most. Males are mildly territorial, spacing themselves out along a shoreline. Females may congregate to oviposit just below the water's surface into well-decayed logs. Nymphs lurk on underwater plants and among bottom detritus. Overwinters as a late-instar nymph.

Powdered Dancer *Argia moesta*

Breeding: Variety of streams and rivers, often rocky; sometimes large lakes.

Nature Notes:

Our most common whitish damselfly, due to the male's extensive pruinosity, and only one on flowing waters (older Lilypad Forktail females also mostly whitish).

Pairs may spend an hour or more underwater while laying eggs!

Often perches on bare ground.

male's abdomen tip; paraprocts unbranched in side view

Description: Adults average 1.6 inches long. A large, light-colored dancer. Male becomes almost completely whitish when mature as pruinosity covers most of his body (inset) except for mid-abdominal segments which remain dark. Immature males are tan to dark brown, turning darker with age before becoming pruinose. Wide, dark shoulder stripes (↑) of male become obscured by pruinosity. Females are brown-form or blue-form, with hairline dark shoulder stripes on thorax (↑). Her abdomen with pale, wide top stripe and narrow dark side stripes, often noticeably pale at tip (↑). Wings of both sexes clear.

Identification Clues: Extensive whitish pruinosity of male is unique. Our only dancer with two cells below the stigma of the wing (↑), separated in the middle by a crossvein (both sexes). Female otherwise very similar to female

This blue-form female shows a very thin dark shoulder stripe. Inset: Note that female has two cells below wing stigma.

Males become covered in a white pruinose bloom as they age.

Blue-fronted Dancer (see *Identification Clues* for that species).

This male's wide shoulder stripe is becoming obscured with pruinosity.

Life Cycle & Behavior: Adults have a long flight period, beginning to emerge in early June and flying into September. Tenerals require two weeks after emergence to mature. More likely than other dancers to hold wings low, along side abdomen. Pairing occurs throughout the summer months. Pair descends, usually in tandem, sometimes deep underwater to oviposit in wet wood or algae-covered rocks and may remain submerged for an hour or more! Nymphs lurk under stones and other bottom debris. Overwinters as a nymph, probably in one of the last instars. Emergence-perches are usually rocks or logs well above water.

Blue-ringed Dancer *Argia sedula*

Breeding: Well-vegetated small to large streams and rivers; occasionally lake shores and ditches.

Nature Notes:

Range recently extended to the northwest with discovery of a population in the Bark River of southeastern Wisconsin.

Often perches higher on vegetation than other dancers.

The scientific species name means "persistent", but the allusion is unknown.

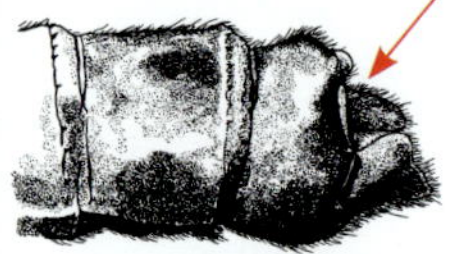

male's abdomen tip; cerci are long and paraprocts show shallow branch in side view

Description: Adults average 1.3 inches long. A small dancer that could easily be mistaken for a bluet. Males show dark blue on thorax with heavy black top and shoulder stripes; those stripes slightly notched at upper end (↑). Blue on thorax becomes lighter on lower sides. His abdomen mostly black with pale blue rings at bases of most segments. Top and sides of last three abdominal segments blue. Eyes and face dark blue. Female completely unlike male, light brown with olive highlights on thorax and abdomen. Her dark top and shoulder stripes very thin. Her abdominal segments with slightly deeper brown apical rings, fading toward lightened tip. Wings of both sexes with very light amber or brown tint.

Identification Clues: Males can be confused with some male American bluets, but none of those species have thick black shoulder stripes

The female is paler in coloration than other female dancers in our region.

notched at upper ends. Male is most like male Stream Bluet, with which it shares habitat, differing in having darker blue pale shoulder stripe, thicker dark shoulder stripe, and three abdomen tip segments all blue on top (Stream Bluet has only segment 9 blue on top). Also, his terminal appendages are unlike any American bluet (pg. 68-69). Female is paler and less patterned than most brown female damselflies in our region, especially at abdomen tip. Amber-tinted wings also help distinguish females.

Life Cycle & Behavior: Adults emerge primarily in June and may fly into September. More likely to perch on streamside vegetation than other dancers, facing the stream. Eggs are laid in tandem into green plant stems at or near waters' surface or into floating vegetation. Overwinters as a late-instar nymph.

Variable Dancer *Argia fumipennis violacea*

MAY	JUNE	JULY	AUG	SEPT	OCT

Breeding: Clear, gentle streams and stream backwaters; also some ponds and lakes.

Nature Notes:

Male is strikingly attractive—our only mostly purple damselfly.

Male assumes "sentinel" position while female oviposits (see lower photo on opposite page).

Flies most actively in morning and during midday.

Sometimes called Violet Dancer, common name of our subspecies, but Variable should be used.

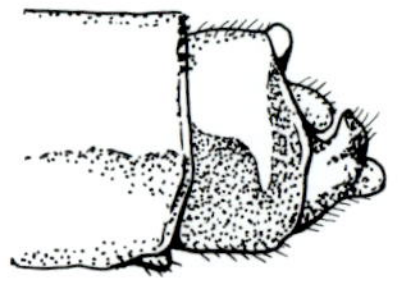

male's abdomen tip

Description: Adults average 1.2 inches long. A small, easily identified, purple dancer. Thorax and abdomen of male is mostly violet, with black, forked shoulder stripes (↑) and narrow black rings on most abdominal segments. Younger male is violet-blue (see inset photo above). Tops of segments 8, 9 and 10 are powder blue. Female is mostly brown with thin, forked shoulder stripes and longitudinal black stripes (↑) on most abdominal segments. Wings of both sexes clear. Two subspecies that occur further south and east have smoky or very dark wings.

Identification Clues: Violet color and forked shoulder stripes of male are unmistakable. Female separable from female Blue-tipped Dancer by her thinner forked shoulder stripe and lighter-colored abdomen on top. Other female dancers lack a forked shoulder stripe. As

With her thinly forked, dark shoulder stripes, the female Violet Dancer is easy to recognize. Also note her longitudinal black stripes atop most abdominal segments.

Two pairs of Variable Dancers ovipositing (laying eggs) with males contact-guarding in the sentinel position.

usual, identity of males with which they are associating is a strong clue to identity of females.

Life Cycle & Behavior: Adults have a long flight period, beginning to emerge in mid June and flying late into September. Perches on ground, rocks, and vegetation. Pair oviposits in tandem on wet logs and other plant materials, occasionally descending below water's surface. Nymphs lurk on plant stems and on other bottom debris. Overwinters as a late-instar nymph.

Springwater Dancer *Argia plana*

MAY	JUNE	JULY	AUG	SEPT	OCT

Breeding: Shallow streams, mid-sized and smaller, including spring and hillside seeps.

Nature Notes:

Once considered a subspecies of Vivid Dancer, a western species of great similarity.

Males holding best territories have greater chances than other males of seizing females as they fly toward breeding sites.

Habitat includes tiny spring seeps and trickles having no other dancers.

male's abdomen tip

Description: Adults average 1.4 inches long. Although males of this jaunty dancer occur in two color forms, blue-form is the only one expected in our region; purple-form occurs much further to the south and west. Thorax and abdomen of male is mostly blue, with a black shoulder stripe unforked (rarely forked) and hairline thin in middle (↑). Top thoracic dark stripe of moderate width. Middle abdominal segments have black apical rings increasing in size away from body and typically have black streaks of variable size on sides (↑). Segment 7 mostly black. Last three segments light bluish. Base color of female can be brown, purple, or blue depending on population. Her markings on thorax similar to male. Her abdomen with black blotches at apices of middle segments and black streaks along the sides. Segment 7 mostly dark. Segments 8-10 entirely pale. Wings of both sexes clear.

♀

Note bright blue abdomen tip of this brown-form female. Mating pair in wheel position (inset).

Identification Clues: Male is the only bright blue dancer in our region with a mostly blue abdomen. Male is distinguished from our blue-type bluets in having an all-blue abdominal segment 10, black streaks on the sides of some mid-abdominal segment, long tibial spurs, and long, forked paraprocts (pg. 68-69). Female differs from female Variable Dancer in having dark shoulder stripe unforked, and from all bluets in having long tibial spurs and black streaks on the sides of her mid-abdominal segments. Not all females will be identifiable in the field, so note identity of males they are associating with.

Life Cycle & Behavior: Adults emerge primarily in June and may fly into September. Males are mildly territorial which serves to space them along shorelines. Female lays eggs in tandem into a wide variety of plant material at or near water surface. Nymphs lurk on plant stems and on other bottom debris. Overwinters as a late-instar nymph.

Blue-tipped Dancer *Argia tibialis*

MAY	JUNE	JULY	AUG	SEPT	OCT

Breeding: Small to mid-sized streams, slow or fast flowing, with banks having some shade. Occasionally sloughs and ponds.

Nature Notes:

Often hovers over water as do Powdered, and sometimes, Blue-ringed dancers.

Prefers shade and is more likely to perch in vegetation than our other dancers.

Look for the Blue-tipped in southerly areas of our region.

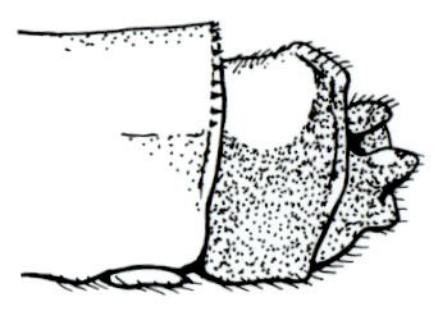

male's abdomen tip

Description: Adults average 1.4 inches long. A smallish, dark dancer. Male is mostly dark, with pale shoulder stripes that are wide and purplish (↑) or brown. Dark shoulder stripes are wide and black. Top of segments 9 and 10 are powder blue. Female's thorax is blue or brown with a wide, dark shoulder stripe forked at top forming a small pale blue triangle (↑). Female's abdomen is dark like male's, but with segment 10 light brown (↑). Wings clear.

Identification Clues: Male is our only dancer with a dark thorax and pale blue abdomen tip. Female is our only dancer with wide, dark shoulder stripe that is forked at the top forming a pale blue triangle. Neither sex likely to be confused with any bluet or forktail.

Life Cycle & Behavior: Adults have a long flight period, beginning to emerge by mid June and flying through August. Female oviposits in

Note her thick dark shoulder stripe with small pale triangle (top). Female Blue-tipped Dancer may show brown (bottom) on the thorax or blue.

wet wood and other plant materials at or slightly above water level, sometimes in aggregations. Perches on ground or low vegetation, but not commonly on rocks. Nymphs lurk under stones and other bottom debris. Overwinters as a nymph, likely in one of the last instars.

Glossary

Antehumeral stripe: Pale stripe on side of the thorax between the dark middorsal stripe and the dark humeral stripe. Commonly called the pale shoulder stripe.

Cerci: (singular: cercus) Upper pair of the terminal appendages (claspers) of the male (a.k.a superior appendages).

Claspers: Common term for the terminal appendages (cerci and paraprocts) of the male that are used to grasp the prothorax of the female during mating (a.k.a. caudal appendages).

Endophytic oviposition: Laying eggs into plant tissue.

Exuviae: Shed skin (exoskeleton) left behind when the nymph climbs out of the water and transforms into the adult (the Latin source word has only the plural form, but many odonatists use a singular form exuvia to avoid ambiguity).

Heterochromatic: Female coloration not resembling the male.

Homeochromatic: Female coloration resembling the male.

Humeral stripe: Dark stripe on side of the thorax below the pale antehumeral stripe. Commonly called the dark shoulder stripe.

Larva: (plural: larvae) Often used name for the immature aquatic stage of the damselfly. Preferably called nymph; sometimes called naiad.

Mesostigmal plates: Small sclerites at the front and top of the pterothorax that are often modified in females for contact with the male claspers when in tandem. Sometimes called "shoulder pads."

Middorsal stripes: Median thorax stripes that are usually dark and flank the middorsal carina on each side.

Nodus: Shallow notch on the leading edge of the wing where the two anterior wings veins (the costa and the subcosta) come together.

Nymph: Immature aquatic stage of the damselfly. Sometimes called larva or naiad.

Occipital bar: Pale, transverse line found along the back of the head of some damselflies between the postocular spots (eyespots).

Ocelli: Three simple eyes found on top of an adult damselfly's head.

Odonatist/Odonatologist: A person who studies dragonflies and damselflies.

Oviposition: The act of laying eggs, which the female may do alone or accompanied by the male.

Ovipositor: The structure on the bottom side of the abdomen at segments 8 and 9 that the female damselfly uses to pierce an appropriate substrate and deposit eggs.

Paraprocts: The lower pair of terminal appendages (claspers) of the male (a.k.a. inferior appendages).

Pharate adult: Newly formed adult that has not yet emerged from the nymphal skin.

Postocular spots: Pale colored spots on top of the head near the eyes, sometimes connected to the occipital bar (a.k.a. eyespots).

Prothorax: The first segment of the thorax, which has no wings but has the first pair of legs.

Pruinosity: A whitish, grayish or pale bluish, waxy substance (bloom) that exudes from the cuticle as some damselflies mature. It may cover much of the body of an older adult (a.k.a. pruinescence).

Pseudostigma: Area of contrasting color at front tip of the wing near where a true stigma would be found, but often larger than a true stigma and containing multiple cells.

Pterothorax: The combined middle (meso-) and hind (meta-) parts of the thorax that are fused together for strength. Bears the wings and two pairs of legs (a.k.a. synthorax).

Stigma: Single cell near the tip of the wing that contrasts in color and texture with the rest of the wing (a.k.a. pterostigma).

Tandem: Position in which the male is holding the thorax of the female with his terminal appendages, but the pair is not copulating.

Teneral: young adult damselfly just after it has emerged from the nymphal stage but before its wings and sclerites are fully hardened and it is full colored (not yet sexually mature).

Vulvar spine: Sharp, pointed extension at the bottom end of abdominal segment 8 on females of some damselflies.

Wheel: The circular or heart-shaped position of mating damselflies when in tandem and with the female engaging the males secondary genitalia with the tip of her abdomen.

Zygoptera: The suborder of Odonata containing the damselflies.

Checklist of Damselflies

MINNESOTA, WISCONSIN & MICHIGAN

BROAD-WINGED DAMSELS *Family Calopterygidae*

Jewelwings

❏ River Jewelwing	*Calopteryx aequabilis*	C
❏ Ebony Jewelwing	*Calopteryx maculata*	A

Rubyspots

❏ American Rubyspot	*Hetaerina americana*	O
❏ Smoky Rubyspot	*Hetaerina titia*	R

SPREADWINGS *Family Lestidae*

Stream Spreadwings

❏ Great Spreadwing	*Archilestes grandis*	U

Pond Spreadwings

❏ Spotted Spreadwing	*Lestes congener*	C
❏ Northern Spreadwing	*Lestes disjunctus*	C
❏ Southern Spreadwing	*Lestes australis*	U
❏ Sweetflag Spreadwing	*Lestes forcipatus*	FC
❏ Lyre-tipped Spreadwing	*Lestes unguiculatus*	O
❏ Slender Spreadwing	*Lestes rectangularis*	C
❏ Emerald Spreadwing	*Lestes dryas*	U
❏ Swamp Spreadwing	*Lestes vigilax*	FC
❏ Elegant Spreadwing	*Lestes inaequalis*	O
❏ Amber-winged Spreadwing	*Lestes eurinus*	O

POND DAMSELS *Family Coenagrionidae*

Eurasian Bluets

❏ Prairie Bluet	*Coenagrion angulatum*	R
❏ Subarctic Bluet	*Coenagrion interrogatum*	R
❏ Taiga Bluet	*Coenagrion resolutum*	FC

American Bluets (Intermediate-type Bluets)

❏ Tule Bluet	*Enallagma carunculatum*	C
❏ Alkali Bluet	*Enallagma clausum*	R
❏ River Bluet	*Enallagma anna*	U

(Black-type Bluets)

❏ Rainbow Bluet	*Enallagma antennatum*	U
❏ Azure Bluet	*Enallagma aspersum*	U
❏ Skimming Bluet	*Enallagma geminatum*	FC
❏ Slender Bluet	*Enallagma traviatum westfalli*	U
❏ Stream Bluet	*Enallagma exsulans*	C
❏ Turquoise Bluet	*Enallagma divagans*	U

(Blue-type Bluets)

❏ Familiar Bluet	*Enallagma civile*	FC
❏ Northern Bluet	*Enallagma annexum*	C
❏ Boreal Bluet	*Enallagma boreale*	C
❏ Marsh Bluet	*Enallagma ebrium*	A
❏ Hagen's Bluet	*Enallagma hageni*	A
❏ Double-striped Bluet	*Enallagma basidens*	U

(Citrus-hued Bluets)

❏ Orange Bluet	*Enallagma signatum*	C
❏ Vesper Bluet	*Enallagma vesperum*	FC

Forktails

❏ Lilypad Forktail	*Ischnura kellicotti*	U
❏ Plains Forktail	*Ischnura damula*	R
❏ Western Forktail	*Ischnura perparva*	VR
❏ Eastern Forktail	*Ischnura verticalis*	A
❏ Fragile Forktail	*Ischnura posita*	U
❏ Citrine Forktail	*Ischnura hastata*	R

Western Red Damsel

❏ Western Red Damsel	*Amphiagrion abbreviatum*	O

Aurora Damsel

❏ Aurora Damsel	*Chromagrion conditum*	FC

Sprites

❏ Sphagnum Sprite	*Nehalennia gracilis*	R
❏ Sedge Sprite	*Nehalennia irene*	A

Dancers

❏ Blue-fronted Dancer	*Argia apicalis*	FC
❏ Powdered Dancer	*Argia moesta*	C
❏ Blue-ringed Dancer	*Argia sedula*	U
❏ Variable Dancer	*Argia fumipennis violacea*	C
❏ Springwater Dancer	*Argia plana*	R
❏ Blue-tipped Dancer	*Argia tibialis*	O

A –	Abundant	Hard to miss. Widespread and abundant.
C –	Common	Encountered often and in many habitats.
FC –	Fairly Common	Fairly easy to find on a given day.
O –	Occasional	May be locally common but not widespread.
U –	Uncommon	Never numerous. Must be in right habitat.
R –	Rare	Right time and right habitat may equal success.
VR –	Very Rare	Strays to our area. Lucky to see one ever.

Damselfly Groups & Websites

Dragonfly Society of the Americas *www.odonatacentral.org*
This is THE website to submit your dragonfly sightings/photos for all of North America. The DSA also publishes the journal *Bulletin of American Odonatology* and the very useful newsletter, *Argia*.

World Dragonfly Association *google "world dragonfly association"*
WDA publishes the international journal *Pantala* and a newsletter, *Agrion*.

Minnesota Dragonfly Society *www.mndragonfly.org*

Wisconsin Dragonfly Society *www.widragonflysociety.org*

Wisconsin Odonata Survey *www.wiatri.net*

Michigan Odonata Survey *google "michigan odonata survey"*
The MOS publishes the newsletter *Williamsonia*.

Ontario Odonata Atlas *google "ontario odonata atlas"*

International Odonata Research Institute *www.iodonata.net*
Comprehensive source for Odonata websites, collecting resources and dragonfly books. Contact information for most of the world's odonatists.

Facebook Group — *Wisconsin Dragonfly Society*

Facebook Group — *Minnesota Dragonfly Society*

Facebook Group — *Michigan Odonata*

Facebook Group — *Ontario Butterflies and Dragonflies*

Facebook Group — *Dragonflies & Damselflies*

Photo Credits

t=top, m=middle, b=bottom

All photos are by Mike Reese except those noted below:

Robert DuBois: 52 inset, 57 bl, 140 inset
Sid Dunkle: 72, 73 t, 75 t, 76 br, 88, 89 t, 91, 112, 116, 122, 123 all
Mike Furtman: 14 B
Kurt Huebner: 38, 39 t, 44, 74 main, 99 b, inset, 104, 105 b, 106, 107 m, 109 t, 136 main, 139 b, 140 main, 141 main, 142, 143 inset, 144
Dan Jackson: 37 top, 50, 51 all, 74 inset, 75 b, 86 main, 98, 99, 102, 103 all, 105 t, 107 t, 108, 113 all, 124, 126 inset, 136 lower inset, 138, 139 t, 141 inset, 145 all
Dennis Paulson: 73 br
Sparky Stensaas (www.ThePhotoNaturalist.com): 1, 7, 15 all, 33, 36, 52, 64, 118, 127, 136 upper inset, 137 t, b
Ken Tennessen: 39 inset, 45, 46, 62, 63, 70, 71, 89 b, 90, 110, 114, 115 all, 117, 133 all, 137 inset, 143 main

Titles of Interest

Acorn, J. 2004. *Damselflies of Alberta: Flying Neon Toothpicks in the Grass.* Edmonton, AB: The University of Alberta Press.

Carpenter, V. 1991. *Dragonflies and Damselflies of Cape Cod.* Cape Cod Museum of Natural History. Natural History Series #4.

Corbet, P. S. 1999. *Dragonflies, Behavior and Ecology of Odonata.* Ithaca, NY: Cornell University Press.

DuBois, B. 2010. *Dragonflies and Damselflies of the Rocky Mountains.* Duluth, MN: Kollath-Stensaas Publishing.

Dunkle, S. W. 1990. *Damselflies of Florida, Bermuda, and the Bahamas.* Gainesville, FL: Scientific Publishers.

Glotzhober, R. C., and D. McShaffrey. 2002. *The Dragonflies and Damselflies of Ohio.* Columbus, OH: Ohio Biological Survey.

Holder, M. 1996. *The Dragonflies and Damselflies of Algonquin Provincial Park.* Algonquin Park Tech. Bulletin #11.

Jones, C. D., A. Kingsley, P. Burke and M. Holder. 2008. *Field Guide to The Dragonflies and Damselflies of Algonquin Provincial Park and the Surrounding Area*, Algonquin Field Guide Series.

Lam, E. 2004. *Damselflies of the Northeast.* Forest Hills, NY: Biodiversity Books.

Mead, K. 2017. *Dragonflies of the North Woods; 3rd Edition.* Duluth, MN: Kollath-Stensaas Publishing.

Needham, J. G., M. J. Westfall and M. L. May. 2000. *Dragonflies of North America.* Gainesville, FL: Scientific Publishers, Inc.

Nikula, B., J. L. Loose, and M. R. Burne. 2003. *A Field Guide to the Dragonflies and Damselflies of Massachusetts.* Westborough, MA: Massachusetts Division of Fisheries and Wildlife.

Paulson, D. 2011. *Dragonflies and Damselflies of the East.* Princeton, NJ: Princeton University Press.

Paulson, D. 2009. *Dragonflies and Damselflies of the West.* Princeton, NJ: Princeton University Press.

Rosche, L. 2002. *Dragonflies and Damselflies of Northeast Ohio.* Cleveland, OH: Cleveland Museum of Natural History.

Walker, E. M., and P. S. Corbet. 1975. *The Odonata of Canada and Alaska, Volume 1.* Toronto, ON: Univ. Toronto Press.

Westfall, M. J. and M. L. May. 2006. *Damselflies of North America:* 2nd edition. Gainesville, FL: Scientific Publishers.

Index

GHI

JKL

MNO

PQR

S

TUVW

Quick Damselfly Finder